AF541563

Seed Production of Horticulture Crops Principles and Practices

About Authors

M. Jayanthi, did Ph.D (Seed Science and Technology) from Tamil Nadu Agricultural University, Coimbatore. She has published 15 research paper and 5 book chapters. She also attended International Training conducted by International Seed Testing Association at Zurich, Switzerland. Now, she is working as Senior Research Fellow at Tamil Nadu Agricultural University, Coimbatore.

S. Sumathi, completed her M.Sc.(Ag), Ph.D. in Seed Science and Technology from Tamil Nadu Agricultural University, Coimbatore, India. At present she is working as Research Associate in same university. She has published 15 research papers in national and international journals and authored in 10 book chapters.

B. Venudevan, did Ph.D (Seed Science and Technology) from TNAU, Coimbatore. He is a Recipient of the UGC fellowship for Ph.D thesis work. Published 12 research papers authored 3 books and 11 book chapters. Specialized in medicinal plant seed production. Presently working as Assistant Professor (SST), Department of Plant Breeding and Genetics, TNAU, Killikulam.

Seed Production of Horticulture Crops Principles and Practices

M. Jayanthi
S. Sumathi
B. Venudevan

NEW INDIA PUBLISHING AGENCY
Pitam Pura, New Delhi – 110 088

NEW INDIA PUBLISHING AGENCY
101, Vikas Surya Plaza, CU Block, LSC Market
Pitam Pura, New Delhi 110 034, India
Phone: + 91 (11) 27 34 17 17 Fax: + 91 (11) 27 34 16 16
Email: info@nipabooks.com
Web: www.nipabooks.com

Feedback at feedbacks@nipabooks.com

ISBN: 978-93-87973-32-9

Composed and Designed by NIPA

Preface

The book has been prepared keeping in view the course requirements of seed science and technology for undergraduate students at various institutions. An attempt has been made to consolidate the scattered information and presented in a simple format. The subject matter covered is rather wide and has been discussed in different topics. From this the readers will get knowledge about seeds, science and technology involved in this subject. Seeds are fertilised mature ovules shaped through sexual reproduction in plants. It is the cheapest and key input in agriculture. It is estimated that good quality seeds of improved varieties can contribute about 20-25% increase in yield. Seed technology is an interdisciplinary science, involves such activities as variety development, evaluation and release seed development, seed processing, seed storage, seed testing, seed certification, seed quality control and seed marketing etc., through which the genetic and physical characteristic of seeds could be improved. Each topic was discussed in separate chapter.

We hope that this book will be helpful to our readers of this subject.

Authors

Contents

1

Introduction to Seed and Seed Quality

Seed

- In broad sense, seed is a material which is used for planting or regeneration purpose. However, scientifically, seed is a fertilized matured ovule together covered with seed coat is called seed or it is a propagating material *i.e.*, part of agriculture, sericulture, silviculture and horticultural plants used for sowing or planting purpose.
- Seed is a mature integumented megasporangium or mature ovule consisting of embryonic plants together with store food material covered by a protective coat.
- Seeds are fertilised, mature ovules, the result of sexual reproduction in plants. Not all plants produce seed (some produce spores or reproduce vegetatively) but most of the higher plants produce seed.
- It is the product of the ripened ovule of gymnosperm and angiosperm plants which occurs after fertilization and some growth within the mother plant.
- The formation of the seed completes the process of reproduction in plants (started with the development of flowers and pollination), with the embryo developed from the zygote and the seed coat from the integuments of the ovule.

Importance of seed

- Seed is a vital input in crop production;
    - It is the cheapest input in crop production and key to agriculture progress.
    - Crop status largely depends on the seed materials used for sowing
    - Response of other inputs in crop production depends on seed material used

- The seed required for raising crop is quite small and its cost is so less compared to other inputs
- This emphasis the need for increasing the areas under quality seed production
- It is estimated that good quality seeds to improved varieties can contribute about 20-25% increase in yield.

Seed Technology

- The discipline of study having to do with seed production, maintenance, quantity and preservation
- Seed technology as the method through which the genetic and physical characteristic of seeds could be improved. It is an interdisciplinary science, involves such activities as variety development, evaluation and release seed production, seed processing, seed storage, seed testing, seed certification, seed quality control and seed marketing etc.

Role of seed technology

- A carrier of new technologies
- A basic tool of secured food supply
- The principal means to secure crop yields in less favourable production areas
- A medium and rapid rehabilitation of agriculture in cases of natural disaster

Goals of seed technology

- Rapid multiplication
- Timely supply
- Assured high quality of seed
- Reasonable price

What is seed quality

Seed quality is a relative term and means the degree of excellence when compared to an acceptable standard. The capacity of the seeds is fully exerted only when it possess its own quality in terms of physical, physiological, genetic and health aspects. The seeds having required standards of purity, germination and other attributes are referred as quality seeds.

Only quality seed can reproduce well and preserve the future generation without any deviation in original characters. The availability of quality seeds in time and at affordable price is prime factor which enable the growers to produce uniform, healthy and vigorous seedlings and greater yield.

Characters of quality seed

a. Genetic purity

Genetic purity of seeds refers to the trueness to type. If the seed possesses all the genetic qualities that breeder has placed in the variety, it is said to be genetically pure. It has direct effect on ultimate yields. If there is any deterioration in the genetic makeup of the variety during seed multiplication and distribution cycle, there would definitely be proportionate decrease in its performance. It is, therefore, necessary to ensure genetic purity during production cycles. The genetic purity of seed classes as follows

- Breeder/nucleus : 100 %
- Foundation seeds : 99.5%
- Certified seeds varieties : 98.0%
- Certified hybrid seeds : 95.0%
- Certified hybrid cotton : 90.0%
- Certified hybrid castor : 85.0%

b. Physical purity

It refers to cleanliness of the seed. Seed should be pure, homogeneous and free from inert matter, other crop seed and weed seed. This is needed to maintain seed health and to maintain the genetic and physiological status of seed. Higher the content of pure seed the better would be the seed quality. It determines the planting value of the seed. The physical purity of crop seed as follows

- Bhendi : 99%
- All crops : 98%
- Ragi, sesame, soybean, jute : 97%
- Ground nut : 96%
- Carrot : 95%

c. Physiological quality

Seed germination and vigour and its maintenance of higher order is known as the physiological status of seed. It is needed to maintain the planting value of seed.

d. Seed health

Healthy seed should be free from insect and pathogen attack and with lesser deterioration scale. It is needed to maintain the other mentioned qualities of seed.

Other seed quality characters

- Seed moisture content
    - Cereals : 10-12%
    - Pulses : 7-9%
    - Oilseeds : 6-7%
    - Vegetables : 5-6%
- Seed maturation
- Seed size
- Seed colour

Benefits of using quality seeds

- They are vigorous and genetically pure (true to type)
- The good quality seed has high return per unit area as the genetic potentiality of the crop can be fully exploited
- Less problem of disease, insect, weed seed/other crop seeds
- Minimization of seed/seedling rate *i.e.*, fast and uniform emergence of seedling and maturity
- They can be adopted themselves for extreme climatic condition and cropping system of the location
- The quality seed respond well to the applied fertilizers and nutrients
- Crop raised with quality seed are aesthetically pleasing
- Good seed prolongs life of a variety
- Yield prediction is very easy

- Handling in post-harvest operation will be easy
- Preparations of finished products are also better

High produce value and their marketability

The differences between seed and grain

S.No.	Seed	Grain
1.	It is the result of well planned seed programme	It is the part of commercial produce, saved for sowing/ planting purpose
2.	It is the result of sound scientific knowledge, organized effort, investment on processing, storage and marketing facilities	No such knowledge or effort is required
3.	The pedigree of the seed is ensured. It can be related to the initial breeders seed	Its varietal purity is unknown
4.	During production effort is made to rogue out off types, diseased plants, objectionable weeds and other crop plants at appropriate stages of crop growth which ensures satisfactory seed purity and health	No such effort is made. Hence, the purity and health status may be inferior
5.	The seed is scientifically processed, treated and packed and labelled with proper lot identity	The grain used as seed may be manually cleaned. In some cases, prior to sowing it may also be treated. This is not labelled.
6.	The seed is tested for planting quality namely germination, purity admixture of weed seeds and other crop seeds, seed health and seed moisture content	Routine seed testing is not done
7.	The seed quality is usually supervised by an agency not related with production (Seed certification agency)	There is no quality control
8.	The seed has to essentially meet the quality standards. The quality is therefore well known	No such standards apply here. The quality is non descript and not known
9.	The labels/ certification tags on the seed containers serves as quality marks	No such standards apply here.

2

History of Seed Industry in India

History and development of seed Industry in India can be discussed under two heads

1. Pre-independence development.
2. Post-independence development.

1. Pre-independence Development

During early years of 20th century efforts were made to develop improved varieties of cotton, wheat, groundnut and sugarcane. The state department of agriculture adopted two methods for multiplication and distribution of these improved varieties

- The seed of improved varieties were multiplied at one location and distributed over a large area to replace local varieties.
- The seed was distributed in small packets to large number of farmers and it was expected that farmers would multiply their own seed.

In 1925 Royal Commission on Agriculture was constituted and it made the following recommendations for introduction and spread of improved varieties.

- Separate organization should be there within agriculture to deal with seed testing and distribution.
- The seed distribution should be made through co-operative societies, department of agriculture and seed growers.
- Private seed growers should be given encouragement.

Following the suggestions of Royal commission, the Government of India established several Research Institutes, however the seed multiplication and distribution was not encouraged. Later on several committees were made, notable among them are; Sir John Russell Committee in 1937, ICAR committee in 1940, Dr. Burns committee in 1944, Famine enquiry committee in 1944 and Food Grain Policy committee in 1944. These review commissions revealed that:

- Crop botanist should be involved in development of improved varieties, their testing and demonstrations
- Initial seed should be multiplied on government farms and subsequent multiplication in the fields of registered growers
- Government should purchase the seed from registered growers at a premium price and later on distribute to farmers at concessional rates.
- Till 1939 vegetable seeds were imported from other countries and due to IInd world war in 1939, the supply of vegetable seeds were stopped. By 1945 private vegetable seed companies had developed facilities for producing vegetable seed at Quetta in Pakistan and Kashmir valley. In 1946, All India Seed Growers, Merchants and Nurserymen's Association was formed with the objective of rapid development of vegetable seed industry.

2. Post Independence Development

- 1948-49 - Dr S. Radha Krishnan recommended the formation of Agricultural Universities and he called them Rural Universities.
- 1950 - Recommendations were given by experts from foreign countries for the establishment of Agricultural Universities.
- Until 1951 agriculture was neglected and after the formation of Agricultural University and Research Institutes, agriculture development started in India.

First Five Year Plan (1951-56)

In 1952 - Grow More Food Enquiry Committee was constituted they recommended seed multiplication and distribution schemes. However the progress, made during this period was poor. The schemes proved useful in following manner:

- The system of distribution of improved seeds of food grains came into existence.
- The experience gained in the operation of these schemes was helpful in coordinating agricultural work.

Second Five Year Plan (1956-61)

- In 1957, All India Coordinated Maize Improvement Project was started in collaboration with Rockfeller foundation of USA with multidisciplinary approach. Within four years of its establishment four maize hybrids were released and similar projects were started in Sorghum and Bajra in 1960 and the first sorghum hybrid was released in 1964 and bajra hybrid in 1965. Later on it extended to all crops of economic importance.

- During this period importance was given for multiplication of foundation seed from breeder seed at block level. For this a policy was made that each National Extension Service Block should have a seed farm and a seed store.
- In 1959 first Indo-American Agriculture Production Team was constituted to examine India's food production problems. The team headed by Dr. Sherman E.Johnson of Ford Foundation made following suggestions.
    - Village, Block and district level extension workers should be made responsible for educating farmers in use of improved seed
    - State Agricultural Department should be made responsible for seed certification
    - Co-operatives and private seed growers should be made responsible for seed supply
    - Setting up of seed testing laboratories and
    - Development of uniform seed certification standards and seed laws.

Inspite of significant developments made, the desired progress could not be achieved during this period due to following reasons

- Requisite quantities of breeder seed were not available. Only hybrid maize, bajra were included in seed programs
- Foundation seed production at block level was not organized properly and produced foundation seed was not fully utilized
- Timely inspections were not made
- Marketing of improved seed was largely left to seed producers
- The seed procurement was unsatisfactory for want of adequate funds
- Seed processing was defective and there were large number of complaints regarding purity and germination of seeds

Third Five Year Plan (1961-66)

Central Seeds Corporation (National Seeds Corporation, 1963) was established with the aim of :

1. To establish foundation and certified seed corporation.
2. To assist in the development of seed production and marketing of seeds.
3. To encourage and assist in development of seed certification programs, seed law and seed law enforcement programs.

4. To train personnel involved in seed programs and
5. Co-ordinate the improved seed programs.

Initially NSC was established for foundation seed production but as there was no other organization, NSC has the taken the responsibility of foundation and certified seed production, seed certification and seed marketing. In addition to these it also assisted in setting up of seed processing plants, helping private seed producers and training individuals in seed production programs.

Annual Plans (1966-69)

Notable developments during this period were enactment and enforcement of seed legislation, review of seed status by a seed review team constituted by the government of India in 1968, some of the recommendations were :

- Compulsory registration of varieties marketed as seeds.
- Elimination of varieties of doubtful value.
- Pre-release publicity to be avoided and arrangement should be made for pre-release multiplication of promising varieties.
- Persons/institutions in plant breeding research are required to register with ICAR.
- NSC should continue as national agency for foundation seed production and Agricultural Universities should also be developed for this purpose.
- Co-operative and private seed growers should be encouraged in seed production, processing and marketing. Government agencies should concentrate on planning, research and extension.
- NSC should assist state government in setting up of seed certification agencies and transfer its certification work to them.
- ICAR should lay down the standards to improve the quality of breeder seed.

Fourth Five Year Plan (1969-74)

Tarai Development Corporation was established in 1969 with the assistance from World Bank. The project aimed at integrated agriculture development of Tarai area with the production of quality seed as primary objective. Later on it has been renamed as U.P. Seeds and Tarai Development Corporation Ltd. The unique features of this corporation were

- Involvement of G.B. Pant University of Agriculture and Technology
- Integrated development approach
- Participation of seed growers as shareholders of the corporation in contrast to the contract system of seed production
- Compact Area Approach
- Strict Quality control
- Money Back Guarantee
- Integrated approach for marketing seeds

In 1971 - Indian society of Seed Technology was formed. The society provides opportunities for exchange of experiences and scientific knowledge to persons engaged in seed industry.

Fifth Five Year Plan (1974-77)

National Commission on Agriculture carried out a review of seed industry and it gave its recommendations in 1976, which are as follows :

- Seed industry should be expanded on commercial lines
- ICAR and the Central Seeds Committee should develop a system of national registry of varieties.
- Encouragement should be given to small participants to form compact areas for seed production.
- Promotional measures should be given for seed growers such as seed crop insurance, exemption of taxes etc.
- Development and fabrication of seed processing equipment. Seed processing should be made compulsory.
- Storage of breeder seed and nucleus seed should be done under controlled condition.
- Grow out test should be made an integral part of seed testing.
- Rigorous enforcement of seed act.
- Compulsory certification may be desirable for seed material of hybrids and vegetatively propagated crops.
- Teaching of seed production technology should be introduced in Agricultural Universities.

- Department of agriculture should have three distinct wings dealing with :

  1. Input aspects 2. Law enforcement 3. Seed certification

Based on the recommendations of NCA, Government of India decided to establish seed production agencies in the country for assured supply of seed for increasing agricultural production. The government of India in late 1974 decided to launch National Seeds Programme with the assistance of World Bank. NSP-I was implemented in 1975-76, the actual production started in 1976. During the first phase SSC were established in four states namely Punjab, Haryana, Maharashtra and Andhra Pradesh. The programme was further expanded during phase II and SSC were established in another five states namely Karnataka, Rajasthan, U.P., Bihar and Orissa.

Sixth Five Year Plan (1980-85)

The seed production and distribution net work was further expanded during this period. Seed control order was passed declaring seeds as an essential commodity.

Seventh Five Year Plan (1985-90)

- During this period emphasis was given for infrastructure development and facilities for enhancing seed production both in public and private sector.
- Under the NSP-III phase, State Seed Corporation was established in another four states namely, Assam, West Bengal, M.P. and Gujarat.
- Strengthened the seed technology research and training facilities.

Eighth Five Year Plan (1992-97)

Increased seed production targets have been fixed for the 8th plan period. The renewed emphasis on hybrid seed production is likely to give new dimensions to the Indian seed industry.

Ninth Five Year Plan (1997-2002)

- Ninth Plan focused more on development of air-conditioned storage for breeder and foundation seeds, aerated storage for certified seeds, mobile facilities or common facilities for processing, drying etc. for all seed producing clusters in the country.
- Comprehensive amendments to prevailing Seed Acts, will be required to be enacted.
- Created the National Seed Grid.

- Special monitoring arrangements for production and lifting of breeder seeds by users, storage of these seeds, scientific assessment of seed requirements of various varieties of different crops for different areas and States, etc.
- Distribution of minikits under various programmes, would be enlarged and more areas will be brought under this programmes. Also, frontline demonstrations would be carried out covering larger areas. Emphasis will be laid on promoting higher seed replacement rate. The most important of all would be the coverage of drought-prone, difficult, hilly and other non-accessible areas under various programmes of distribution of seeds. Care will also be taken so that an effective system of import/export of certified seeds is developed.
- Efforts will be made to have effective linkages between the public and the private sectors where both are mutually supporting and complementary in their activities. The seed policy will encourage the private sector to bring its production under the purview of certification or other accepted systems of quality control.

Tenth Five Year Plan (2002–2007)

- There is a large increase in the distribution of certified seeds.
- Private sector has expanded to fill the gap. The private sector seed industry in India is growing appreciably and has made significant contributions to BT cotton, hybrids of maize, rice, sunflower, etc.

Eleventh Five Year Plan (2007-2012)

- It is high time that the private sector is invited to participate in the large-scale multiplication of breeder seeds into certified seeds so that replacement rates can be significantly increased.
- An effort to raise the seed replacement rate was redoubled, as this is the foundation for accelerating productivity growth.
- There is also an urgent need to check the supply of spurious seeds by many companies by improving the governance of regulatory bodies, and also by keeping an eye on the monopoly practices of seed companies.

Tweleth Five Year Plan (2012–2017)

- Supply of region specific varietal seeds and ensuring adequate stocks and timely supply of quality seeds to farmers for which the State will prepare a comprehensive Seed Plan.

- Sensitizing farmers on production and usage of indigenous and certified seeds.
- Encouraging private entrepreneurs in quality seed production.
- Involving farmers, women SHGs, TANWABE groups and NGOs in local seed production and seed processing activities.
- Improving infrastructure facilities for seed production, processing and storage Enhancing Seed Replacement Rate (SRR).

An Action Plan for all the blocks has been prepared by incorporating details of crop wise/variety wise requirement of breeder seeds, foundation seeds and certified seeds for the next five years.

3

Generation System of Seed Multiplication

Generation system of seed multiplication is nothing but the production of a particular class of seed from specific class of seed up to certified seed stage. The choice of a proper seed multiplication model is the key to further success of a seed programme. This basically depends upon,

- The rate of genetic deterioration
- Seed multiplication ratio
- Total seed demand (Seed replacement rate)

Based on these factors different seed multiplication models may be derived for each crop and the seed multiplication agency should decide how quickly the farmers can be supplied with the seed of newly released varieties, after the nucleus seed stock has been handed over to the concerned agency, so that it may replace the old varieties. In view of the basic factors, the chain of seed multiplication models could be,

- Three generation model: Breeder seed - Foundation seed - Certified seed
- Four generation model: Breeder seed - Foundation seed (I)- Foundation seed (II) -Certified seed
- Five generation model: Breeder seed - Foundation seed (I)- Foundation seed (II) - Certified seed (I) - Certified seed (II)

For most of the often cross pollinated crops three and four generation models is usually suggested for seed multiplication, *e.g.* Castor, Red gram, Jute, Green gram, Rape seed, Mustard, Sesame, Sunflower & most of the vegetable crops.

Classes of seed

Nucleus seed

Nucleus seedis the handful of original seed obtained from selected individual plants of a particular variety for maintenance and purification by the originating breeder. It is further multiplied and maintained under the supervision of qualified plant breeder to provide breeder seed. This forms the basis for all further seed production. It has the highest genetic purity and physical purity.

Breeder seed

This is the progeny of the nucleus seed multiplied in large area under the supervision of plant breeder in a research institute/ Agricultural University and monitored by a committee. It provides cent per cent physical and genetic pure seed for production of foundation class. Golden yellow coloured certificate is issued for this category by the producing agency.

Foundation seed

It is the progeny of breeder's seed in handled by recognized seed producing agencies like State agricultural universities, State farm corporation of India, National Seed Corporation, State seed corporation and private sectors under the supervision of Seed Certification Agency in such a way that its quality is maintained according to the prescribed standard. Seed Certification agency issues a white colour certification for foundation class seed. Foundation seed is purchased by Seed Corporation from seed growers. Foundation seed can again be multiplied by Seed Corporation in the events of its shortage with similar seed certification standard.

Registered seed

It is the progeny of foundation seed so handled as to maintain its genetic identity and purity and approved and certified by a certifying agency. It should be of quality suitable to produce certified seed.

Certified seed

Progeny of foundation seed produced by registered seed growers under the supervision of Seed Certification Agency by maintaining the seed quality as per minimum seed certification standards. Seed Certification Agency issues a blue colour certificate.

Differences between certified seed and truthful seed

Certified seed	Truthful labelled seed
Certification is voluntary	Truthful labelling is compulsory for notified kind of varieties
Applicable to notified kinds only	Applicable to both notified and released varieties
It should satisfy minimum field and seed standards	Tested for physical purity and germination
Seed certification officer, seed inspectors can take samples for inspection	Seed inspectors alone can take samples for checking

Seed multiplication ratio (SMR)

It is nothing but the number of seeds to be produced from a single seed when it is sown and harvested. The seed multiplication ratio of different crops is as follows.

Crops	SMR
Okra	1:100
Tomato	1:400
Brinjal	1:450
Chillies	1:240
Watermelon	1:100
Pumpkin	1:160
Bitter gourd	1:41
Bottle gourd	1:99
Ridge gourd	1:83
Cucumber	1:200
French bean	1:9
Cluster bean	1:50
Peas	1:19
Onion	1:171
Radish	1:100
Carrot	1:83

Seed Replacement Rate (SRR)

Seed Replacement Rate is the rate at which the farmers replace the seeds with quality seeds instead of using their own seeds. Seed Replacement Rate of major vegetables.

Crops	SRR (%)
Brinjal	63.4
Cabbage	100
Cauliflower	86.4
Chilli	83.7
Gourds	73.5
Melons	89.2
Okra	92.4
Tomato	99.3
Beans	62.2
Onion	87.3
Peas	93.5
Others	72.6

4

Varietal Deterioration: Causes and Maintenance

Variety is a group of plants having clear distinguished characters which when reproduced either sexually or asexually retains their characters. The genetic purity of a variety or trueness to its type deteriorates due to several factors during the production cycles. The factors that are responsible for loss of genetic purity during seed production are :

- Developmental Variation
- Mechanical Mixtures
- Mutations
- Natural Crossing
- Genetic drift
- Minor Genetic Variation
- Selective influence of Diseases
- Techniques of the Breeder
- Breakdown of male sterility
- Improper/defective seed certification System

Developmental Variation

Seed crop is grown in difficult environmental conditions such as different soil and fertility conditions, under saline or alkaline conditions or under different photo-periods or different elevations or different stress conditions for several consecutive generations the developmental variations may arise as differential growth response.

To avoid or minimize such developmental variations the variety should always be grown in adaptable area or in the area for which it has been released. If due to some reasons (for lack of isolation or to avoid soil born diseases) it is grown in non adaptable areas it should be restricted to one or two seasons and the basic seed *i.e.* nucleus and breeder seed should be multiplied in adaptable areas.

Mechanical Mixtures

This is the major source of contamination of the variety during seed production. Mechanical mixtures may take place right from sowing to harvesting and processing in different ways such as

- Contamination through field – self sown seed or volunteer plants
- Seed drill – if same seed drill is used for sowing 2 or 3 varieties
- Carrying 2 different varieties adjacent to each other
- Growing 2 different varieties adjacent to each other.
- Threshing floor
- Combine or threshers
- Bags or seed bins
- During seed processing

To avoid this sort of mechanical contamination it would be necessary to rogue the seed fields at different stages of crop growth and to take utmost during seed production, harvesting, threshing, processing etc.

Mutations

It is not of much importance as the occurrence of spontaneous mutations is very low. If any visible mutations are observed they should be removed by rouging. In case of vegetatively propagated crops periodic increase of true to type stock would eliminate the mutants.

Natural Crossing

It is an important source of contamination in sexually propagated crops due to introgression of genes from unrelated stocks/genotypes. The extent of contamination depends upon the amount of natural cross-fertilization, which is due to natural crossing with undesirable types, offtypes and diseased plants. On the other hand natural crossing is main source of contamination in cross-fertilized or often cross-fertilized crops. The extent of genetic contamination in seed fields is due to natural crossing depends on breeding system of the species, isolation distance, varietal mass and pollinating agent. To overcome the problem of natural crossing isolation distance has to be maintained. Increase in isolation distance decreases the extent of contamination. The extent of contamination depends on the direction of the wind flow, number of insects presents and their activity.

Genetic Drift

When seed is multiplied in large areas only small quantities of seed is taken and preserved for the next years sowing. Because of such sub-sampling all the genotypes will not be represented in the next generation and leads to change in genetic composition. This is called as genetic drift.

Minor Genetic variation

It is not of much importance, however some minor genetic changes may occur during production cycles due to difference in environment. Due to these changes the yields may be affected. To avoid such minor genetic variations periodic testing of the varieties must be done from breeder's seed and nucleus seed in self-pollinated crops Minor genetic variation is a common feature in often cross-pollinated species; therefore care should be taken during maintenance of nucleus and breeder seed.

Selective influence of Disease

Proper plant protection measures much be taken against major pests and diseases otherwise the plant as well as the seeds get infected.

- In case of foliar diseases the size of the seed gets affected due to poor supply of carbohydrates from infected photosynthetic tissue.
- In case of seed and soil borne diseases like downy mildew and ergot of Jowar, smut of bajra and bunt of wheat, it is dangerous to use seeds for commercial purpose once the crop gets infected.
- New crop varieties may often become susceptible to new races of diseases are out of seed production programms, *e.g.* Eg. Surekha and Phalguna became susceptible to gall midge biotype.

Techniques of the Breeder

Instability may occur in a variety due to genetic irregularities if it is not properly assessed at the time of release. Premature release of a variety, which has been breed for particular disease, leads to the production of resistant and susceptible plants which may be an important cause of deterioration. When Sonalika and

Kalyansona wheat varieties were released in India for commercial cultivation the genetic variability in both the varieties was still in flowering stage and several secondary selections were made by the breeders.

Breakdown of male sterility

Generally in hybrid seed production if there is any breakdown of male sterility in may lead to a mixture of F1 hybrids and selfers.

Improper Seed Certification

It is not a factor that deteriorates the crops varieties, but is there is any lacuna in any of the above factors and if it has not been checked it may lead to deterioration of crop varieties.

Maintenance of genetic purity

Horne (1953) had suggested the following methods for maintenance of genetic purity :

- Use of approved seed in seed multiplication
- Inspection of seed fields prior to planting
- Field inspection and approval of the Crop at critical stages for verification of genetic purity, detection of mixtures, weeds and seed borne diseases.
- Sampling and sealing of cleaned lots.
- Growing of samples with authentic stocks or Grow -out test.

Various steps suggested by Hartman and Kestar (1968) for maintaining genetic purity are as follows

- Providing isolation to prevent cross fertilization or mechanical mixtures
- Rouging of seed fields prior to planting
- Periodic testing of varieties for genetic purity
- Grow in adapted areas only to avoid genetic shifts in the variety
- Certification of seed crops to maintain genetic purity and quality
- Adopting generation system

Safe guards for maintenance of genetic purity

The important safe guards for maintaining genetic purity during seed production are

- Control of seed source
- Preceding crop requirement
- Isolation
- Rouging of seed fields
- Seed certification
- Grow out test

Control of Seed Source: The seed used should be of appropriate class from the approved source for raising a seed crop.

Preceding Crop requirement: This has been fixed to avoid contamination through volunteer plants and also the soil borne diseases.

Isolation: Isolation is required to avoid natural crossing with other undesirable types, off types in the fields and mechanical mixtures at the time of sowing, threshing, processing and contamination due to seed borne diseases from nearby fields. Protection from these sources of contamination is necessary for maintaining genetic purity and good quality of seed.

Rouging of Seed Fields: The existence of off type plants is another source of genetic contamination. Off type plants differing in their characteristics from that of the seed crop are called as off types. Removal of off types is referred to as roughing. The main sources of off types are

- Segregation of plants for certain characters or mutations
- Volunteer plants from previous crops or
- Accidentally planted seeds of other variety
- Diseased plants

Off type plants should be rouged out from the seed plots before they shed pollen and pollination occurs. To accomplish this regular supervision of trained personnel is required.

Seed Certification: Genetic purity in seed productions maintained through a system of seed certification. The main objective of seed certification is to make available seeds of good quality to farmers. To achieve this qualified and trained personnel from SCA carry out field inspections at appropriate stages of crop growth. They also make seed inspection by drawing samples from seed lots after processing. The SCA verifies for both filed and seed standards and the seed lot must confirm to get approval as certified seed.

Grow-out Test: varieties that are grown for seed production should be periodically tested for genetic purity by conducting GOT to make sure that they are being maintained in true form. GOT test is compulsory for hybrids produced by manual emasculation and pollination and for testing the purity of parental lines used in hybrid seed production.

5

Methods and Tools for Variety and Hybrid Seed Production

Seed Production

Systemized crop production is known as seed production. In seed production adequate care is given from the purchase of seeds upto harvest adopting proper seed and crop management techniques. The benefits of seed production are

- Higher income
- Higher quality seed for next sowing

Difference between seed and crop production

Seed production	Crop production
Basic seed should be from an authentic source	Any seed material can be used
Seed plot should be selected carefully for better performance, as per edaphic and environmental requirement	Can be grown in any area
Needs isolation from other varieties	Isolation is not necessary
Needs technical skill for maintenance of quality	Special technical skill is not required
Maintenance of genetic purity is important	Genetic purity is not required
Roguing is compulsorily practiced	Roguing is not practiced
Harvesting should be done at physiological/ harvestable maturity	Harvested at field maturity
Resultant seed should be vigorous and viable	Question of viability does not arise
Importance is given to seed quality rather than the yield	Importance is given more to yield

There are two types (major) of seed production *i.e.*, varietal and hybrid. Seed production based on the type of seed used for multiplication. The difference between varietal and hybrid seed production are as follows

Varietal seed production	Hybrid seed production
It is single parent multiplication	It needs two to many parents
Isolation distance requirement is less	Isolation distance requirement is more
Production is by open pollination	Production is by managed control pollination (Female)
Seed can be used continuously for3/4/5 generations	Seed has to be changed every time
Production technique is uniform (multiplication)	Technique differ with crop
Production care is less	Production care is more
Yield will be lower	Yield will be higher
Profit is less	Profit is higher

Variety seed production

Variety seeds produced through controlled or open pollination. During seed production strict attention must be given to the maintenance of genetic purity and other qualities of seeds in order to exploit the full dividends sought to be obtained by introduction of new superior crop plant varieties. In other words, seed production must be carried out under standardized and well organized conditions.

Hybrid seed production

Hybrid is produced by crossing between two genetically dissimilar parents. Pollen from male parent (Pollen parent) will pollinate, fertilize and set seeds in female (seed parent) to produce F1 hybrid (heterosis) seeds. In self pollinated cross it is difficult to cross but in cross pollinated crops it is easier. The following tools are employed to produce hybrid seeds.

- Hand emasculation and pollination
- Self-incompatibility
- Dicliny : monoecious and dioecious
- Male sterility

Based on the crop behaviour the tools have been adapted for production of hybrid seeds commercially.

1. Hand emasculation and pollination

- Hybrid seeds are produced manually by modifying the plant structure by removal of male organ from female plant before anthesis.

- This system is possible only when the male and female parts of a single flower or plants are separate.
- This is being adopted in bisexual perfect flowers where the androecium is removal with case. By removing the anther column / or male part from female line, the sterility of female line is created and is dusted with the pollen of desired male parent.

2. Self Incompatibility

- Self-incompatibility is a mechanism which avoids self fertilization through recognition of self pollen in or on stigma on the female pistil.
- But when pollen from other plant carried by wind or insects is accepted and sets seeds.
- Self-incompatibility will prevent self pollination (inbreeding) and promotes crosspollination (out breeding) and creates genetic variability.
- SI is seen in hermaphrodite and homomorphic flowers.
- The self-incompatibility response is genetically controlled by one or more multi-allelic loci, and relies on a series of complex cellular interactions between the self-incompatible pollen and pistil.

3. Monoecious and dioecious flowers

Presence of Monoecious and Dioecious type of flowers is also one of the tool for production of hybrid seeds. Here the male sterility is created by manual removal of the male part and crossed with desired male parent. Production of these unisexual / imperfect flowers will naturally leads to out crossing, promotes heterozygosity, genetic variability and genetic exchange.

Monoecious - Flowers are present at different position on the same plant. *e.g.*, Cucumber.

Dioecious- Male and female flowers are in different plant. *e.g.*, Papaya

Sex modification in monoecious and dioecious

Sex expression in dioecious and monoecious plants is genetically determined and can be modified to a considerable extent by environmental and introduced factors such as mineral nutrition, photoperiod, temperature, phytohormones. It helps for production of hybrid seeds.

a. Hormones and chemicals

The morphological differences in various sex types and their specific metabolic characteristics result from the possession of specific patterns of proteins, enzymes and other molecules. Auxin treatments increase the female sex tendency while gibberellins cause a shift towards maleness.

- Femaleness: Auxins- NAA, Etherl, Ethephon, Cytokinis- BA, Brassinosteriods
- Maleness: GA3, AgNO3, ABA Thioporpinic acid, Pthalimide, Paclobutrazol

In gynoecious cucumber there will be increased number of male nodes when sprayed with silvernitrate and gibberlic acid, which made possible for multiplication of gynoecious in hybrid seed production.

b. Environmental influence

Male sex expression of several plant species is favoured by high temperatures and female sex expression by low temperatures.

3. Male Sterility

- Female line that is unable to produce viable pollen is called is male sterility. This male sterile line is therefore unable to self-pollinate and seed formation is dependent upon pollen from the male line.
- In hermaphrodite flowers pollens are non-functional or inactive or sterile while, female gametes functions normally.
- It is the inability of plant to produce or to release functional pollen as a result of failure of formation or development of functional stamens, microspores or gametes.
- Male sterility can be either genetic or cytoplasmic or cytoplasmic-genetic.
- This prevents autogamy and permits crosspollination.
- Sterility is due to nuclear genes or Cytoplasmic gene or both.
- In hybrid seed production process female is a male sterile line crossed with male fertility restorer line to get F_1 hybrid.

a. Cytoplasmic male sterility

Cytoplasmic male sterility (CMS) is caused by the extra nuclear genome (mitochondria or chloroplast) and shows maternal inheritance therefore all the off springs will be male sterile. Manifestation of male sterility in CMS may be either entirely controlled by cytoplasmic factors or by the interaction between cytoplasmic and nuclear factors. In general there are two types of cytoplasms are available: N (normal) and the aberrant S (sterile) cytoplasms, which interferes with the formation of normal pollen grains. CMS lines are available in Onion, flowers and potato.

Disadvantages of CMS lines

1. Insufficient or unstable male sterile
2. Difficulties in restoration system
3. Difficulties with seed production

b. Cytoplasmic genetic male sterility

The sterility is manifested by the influence of both nuclear (Mendelian inheritance) and cytoplasmic (maternally inherited) genes. CGMS systems are widely exploited in crop plants for hybrid breeding due to the convenience to control the sterility expression by manipulating the gene–cytoplasm combinations in any selected genotype.

c. Genetic Male sterility

Male sterility is controlled by mutations in nuclear genes in the single recessive genes (ms) affect stamen and pollen development, but it can be regulated also by dominant genes (MS). MS alleles are generally recessive. A male sterile line is maintained by crossing with heterozygous male fertile line.

Constraint of the use of genetic male sterility

- The maintenance of the male sterile line.

Normally, a GMS line (A-line) is maintained by backcrossing with the heterozygote B-lines (Maintainer lines), but the progeny produced are 50% fertile and 50% male sterile.

6

Seed Development and Maturation

The important events involved in seed development and maturation include

- Pollination
- Fertilization
- Development of the fertilized ovule by cell division
- Accumulation of reserve food material
- Loss of moisture content

Pollination

The mature anthers dehisce and release pollen grains (haploid microspores). Pollen grains are transferred from an anther to the stigma of the flower this process is called as pollination. There are three types of pollination are available. Self-pollination or autogamy means the pollen grains are transferred to stigma of the same flower. If they are transferred to the stigma of another flower means cross-pollination or allogamy. If the cross pollination level is 10-40 % means it is called often cross pollinated crops.

Self-pollination occurs in those plants where bisexual flowers achieve anther dehiscence and stigma receptivity simultaneously called as chasmogamy. The majority of angiosperms bear chasmogamous flowers. In some plants, flowers do not open before pollination such flowers are called cleistogamous, and this is the most efficient floral adaptation for promoting self-pollination.

Cross-pollination is ensured in plants which bear unisexual flowers. In bisexual flowers also self-pollination may be prevented by

a. Self-sterility: inability to produce viable pollen

b. Dichogamy: maturation of male and female organs at different times

c. Herkogamy: where the structure of male and female sex organs proves a barrier to self pollination

d. Heterostyly: where flowers are of different types depending on the length of the style and stigma and pollination occurs only between 2 dissimilar types

e. Selfincompatibility:Inability to viable pollen to fertilize ovule of same flower *e.g.*: Cole crops

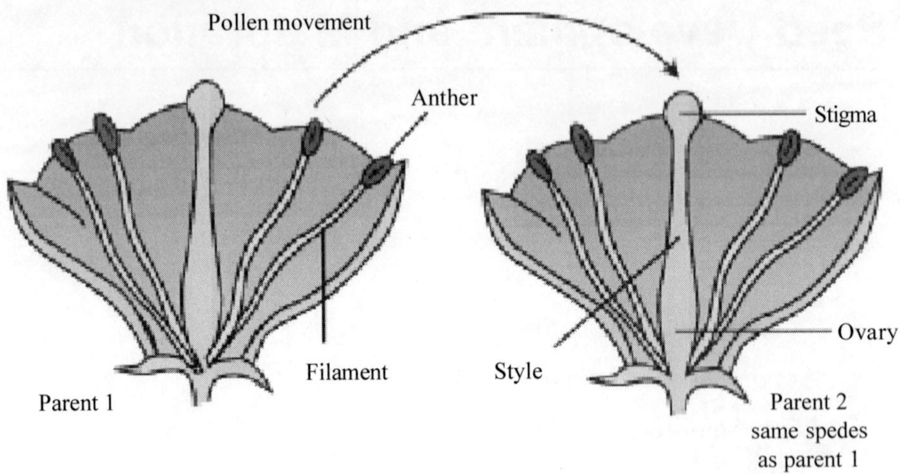

Pollination mechanism

Fertilization

After landing on the stigma, the pollen grain germinates and pollen tube grows through the style. The surface of the stigma secretes substances, which may provide optimum conditions for pollen germination. The pollen tubes traversing the style pectinase which dissolves intercellular substances of the style tissue. After traversing the style, the pollen tube enters embryosac of the ovule. The embryosac consists of 8 cells. The end near the micropyle has the egg apparatus, which consists of egg cell and 2 synergids. There are 2 polar nuclei in the centre and the chalazal end has 3 antipodal cells. In angiosperms, fertilization involves the participation of 2 male nuclei (double fertilization). One fuses with the egg nucleus to form the diploid zygote and the other with 2 polar nuclei to produce a triploid nucleus, which is the primary endosperm nucleus.

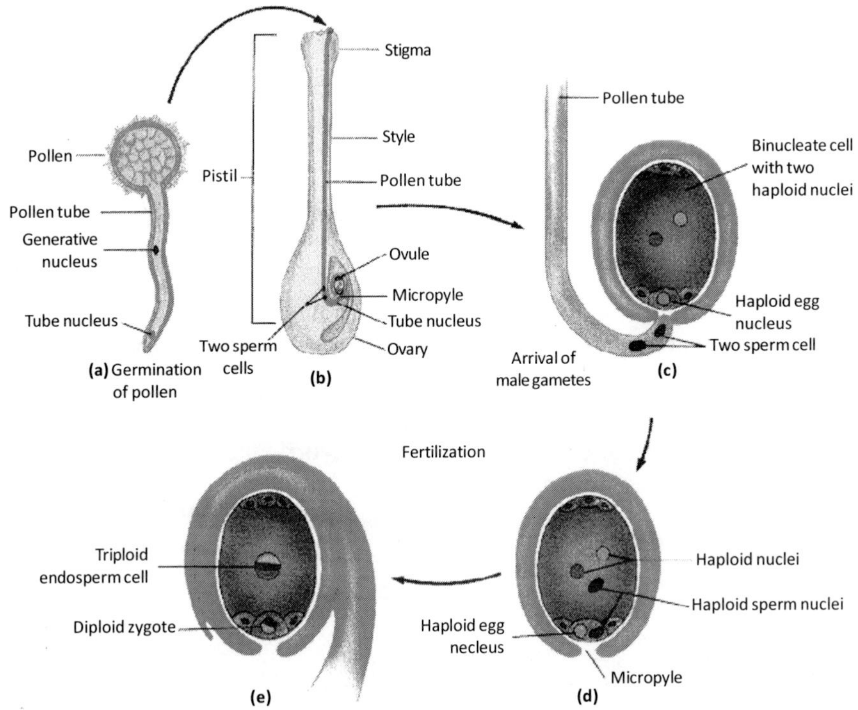

Fertilization mechanism

Seed Development

1. Embryo development

- The first division of the zygote is transverse in dicots and it results in a small apical cell (ca) and a large basal cell (cb).
- Cell ca divides vertically forming 2 juxtaposed cells and cb undergoes a transverse division forming 2 superimposed cells (ci and m). These results in a T-shaped, 4 celled proembryo.
- Cell ci divides transversely giving rise to n and n'. These 2 cells divide further resulting in a row of 3 or 4 cells, forming suspensor.
- Cell m and its derivatives undergo vertical divisions forming a group of 4 to 6 cells. This group divides by oblique-perclinal wall forming a set of inner cells and a row of outer cells. The inner cells form the initials of the root apex and the outer cells form the root cap.
- The 2 cells formed as a result of the division of ca again divide vertically forming quadrant. Each cell of the quadrant divides transversely and thus an octant containing 2 tiers of cell l and p is formed.

- The cells of the octant undergo vertical division resulting in a globular proembryo. Periclinal divisions occur in the peripheral cells of the globular proembryo that delimit an outer layer, the dermatogen.
- The tier l gives rise to cotyledons and shoot apex and l forms hypocotl-radicle axis.
- In monocotyledons, the cell cb remains undivided and develops into a haustorial of the suspension. Cell ca divides into 2 by a transverse division. The terminal cell of these 2 by repeated divisions in different planes gives rise to a single cotyledon. The embryo development in grasses is different from that of other monocotyledons. A dorsiventral symmetry is established as a result of the peculiar oblique position of cell walls early in the embryogeny. The single cotyledon is reduced to absorptive scutellum and additional structures like coleptile and coleorrhiza are formed.

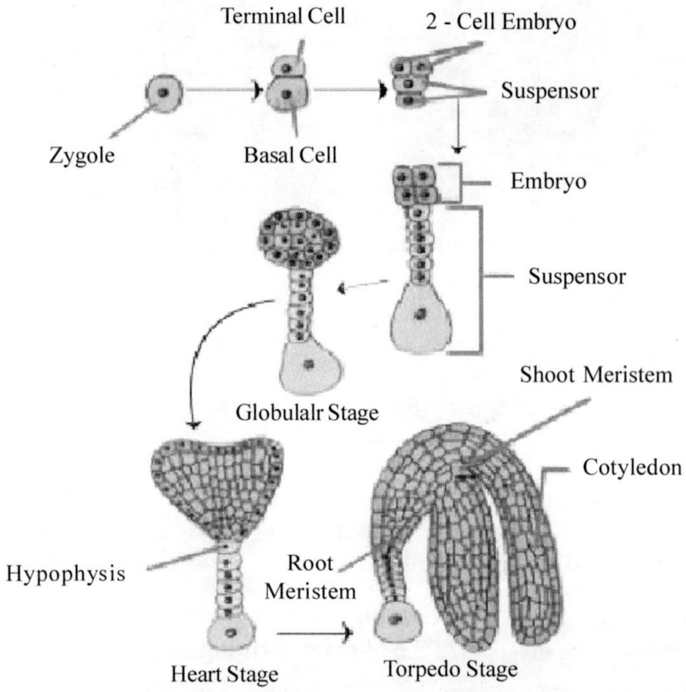

Embryo development in dicots

2. Endosperm development

There are 3 types of endosperm development

a) Nuclear - where the endosperm nucleus undergoes several divisions prior to cell wall formation, *e.g.*, squash.

b) Cellular -in which there is no free nuclear phase, and

c) Helobial where the free nuclear division is preceded, and is followed by cellularization as in some monocots.

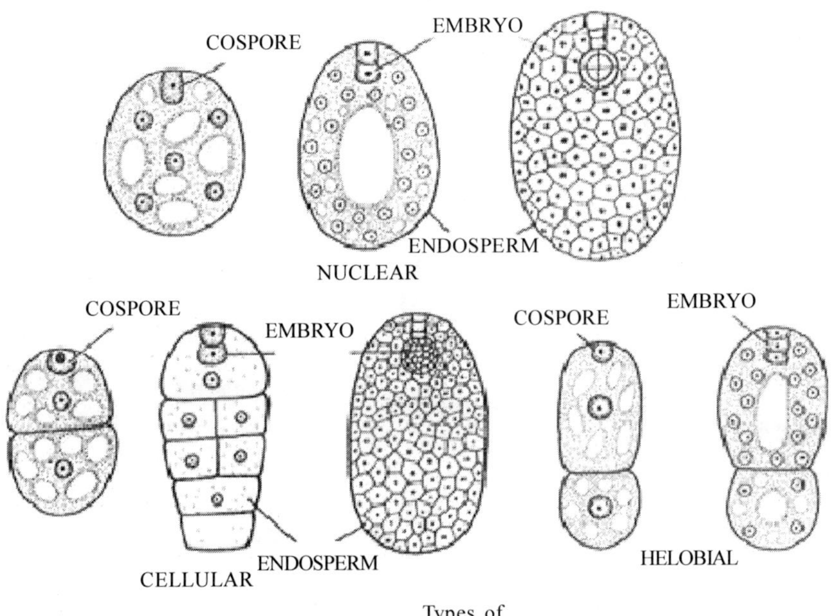

Types of endosperm.

Endosperm development

3. Seed coat development

Integuments of the ovule undergo marked reorganization and histological changes during maturation to form seed coats. In bitegmic ovules (which have 2 integument's), the seed coat may be derived from both the integument's or from the outer integument only, the inner integument may disintegrate.

Maturation

Seed maturation is an important phase of seed development during which embryo growth ceases, storage products accumulate, the protective tegument differentiates and tolerance to desiccation develops, leading to seed dormancy in some crops.

7

Seed Crop Management Techniques

- The seed production of varieties and hybrids should be carried out carefully by following various techniques in the region where they are well adopted.
- The various factors affecting quality seed production includes,
    1. Agro climatic factors / Ecological factors - edaphic and climatic factors
    2. Production factors
    3. Post harvesting handling of seed
    4. Seed quality control factors

1. Agro climatic /Ecological factors

a. Edaphic factors

- Soil should have optimum moisture, good texture and structure. In general, loams are the best. Clay soils in high rainfall area become sick and cause lack of aeration which affects seed quality. Ill drained soils cause chlorosis and wilt diseases.
- Soil pH should be around 7, as neutral. The problematic soil like saline soil and acidic soils are to be avoided.
- Soil should have adequate macro and micro nutrients and microbial load for producing vigorous and viable seeds. For example Boron deficiency causes black rot in cole crops particularly cabbage and cauliflower and hollow heart in garden pea. Molybdenum causes physiological disorder like whiptail in crucifers. Excess nitrogen results more of vegetative growth that leads to more proneness to diseases and insects resulting in reduction of seed quality and yield. N, P and K in balanced dose increases seed yield and improves seed quality and induce resistance.
- The soils should be free from soil borne pathogen, nematodes and weed seeds.

b. Climatic factors

Temperature

- Temperature plays a major role in seed germination, crop growth and maturity. Too high temperature during seed crop maturity brings forced maturity and poor seed quality.
- Optimum temperature is required from sowing to the day of harvest. For *e.g.*, cole crops seed production requires low temperature (4-10°C) at initial stage and high temperature (15-20°C) at reproductive stage *i.e.*, during seed development and maturation.
- Higher temperatures and strong winds cause desiccation of pollen grains and drying of stigma results in poor seed set and seed quality.
- High temperature adversely affects seed production due to drying of anthers in lab-lab; flower shedding in tomato and chillies and production of higher percentage of hard seed in leguminous vegetables. Over wintering (chilling) / vernalization is needed for cabbage, cauliflower, beets, carrots, turnips etc, to have shift from vegetative to reproductive phase which helps in quality seed production.
- In some vegetables, high temperature inhibits development of ovules and fruits and causes shedding of flower buds and young pods / fruits.
- Temperature between 24 - 38°C is most favourable for activities of pollinators particularly bees. Pollinators are an important component in vegetable seed production and without these quality seed production is not possible particularly in cross and often cross-pollinated vegetables. These pollinators stop working at low (below 20° C) or high temperature (beyond 38° C).

Humidity

- Higher relative humidity more than 90% cause heavy flower drop and during maturation will lead to production of blonded seeds (*e.g.*, peas).
- Relative humidity reduced lesser than 40% leads to production of hard seeds.
- Flowering, pollination and seed setting in temperate vegetable needs low humidity and dry weather and moderate to low humidity in sub-temperate and tropical vegetable varieties.
- High humidity and low temperature also encourages production of diseased seeds.

Rainfall

- Excessive rain, apart from affecting pollination, leads to a higher incidence of diseases resulting in mould attack, seed discoloration and storability.
- Activity of pollinators (bees) is practically nil during rains and when flowers are wet.
- It may also results in delayed maturity and at the time of maturation leads to pre-germination or sprouting of seeds in standing crops (eg. peas, beans).
- Strong wind and heavy rainfall at or near harvest may cause heavy seed losses through shattering and also complicate the harvesting operations (*e.g.*, amaranthus).

Cold

- Temperatures below 10°C may not be suitable for tropical crops. It will affect anthesis, pollen germination, pollen fertility delayed growth and maturity, incomplete exertion, reduced filling, choking of panicle and incidence of pest and disease.

Wind

- Wind is necessary for wind pollinated crops. Improves seed setting in highly cross pollinated crops like onion and crucifers.
- At times winds act as a source of contamination.
- Heavy winds may carry pollen too far or prevent deposition on stigma thus reducing seed set.
- Dry winds also desiccate pollen resulting in loss of viability and development of hard seeds in legumes. Heavy winds results in lodging and shattering of seeds/pods.

Insect activity

- Insects are beneficial as well as harmful in seed production.
- Insects acts a source of contamination and in insect pollinated crops one kilometer distance is required as isolation.
- Insects damage seeds right from the pod stage till harvest and account for 20-30% of the seed production losses.

2. Production technology

Selection of suitable production environment based on adoptability

Crop specific temperate, sub tropical and tropical environment should be selected based on their genetic adoptability as it results in better survival capacity of a variety under given environmental condition. For example temperate crops will not set seed in tropical conditions and the virus free production of seed potato is possible only in plains though the multiplication rate is higher at hilly areas.

Selection of land

- Select well levelled, fertile field for uniform maturity.
- Select nearest to irrigation sources.
- Avoid weedy fields particularly the presence of abnoxious weeds.
- Avoid problem fields like alkaline, saline and sodic soils.

Selection of season

- Off season is better to avoid isolation problem.

Selection of crop and varieties

- The variety should be adapted to the agro climatic conditions of the region and having wider adaptability.
- The variety should really be a high yielder and popular one.
- Should possess desirable attributes namely, earliness of the crop, photo and thermal insensitive varieties, drought resistant, pest and disease resistant etc.

Selection of Seeds

- The seed used for raising a seed crop should be of known purity, appropriate class (farmers can produce certified seed).
- Breeder Seed can be produced only by the university and Foundation Seed by State Seed Farm / and invariably obtained from an authorized official agency
- While purchases of seed the details on the tag - agency, purity, germination, validity period should be carefully examined.

Preparation of land

- Perfect leveling and good land preparation helps in improved germination
- Deep ploughing results in destruction of potential weeds, also aids in water management and good uniform irrigation.

Isolation of seed crops

- The seed crop must be isolated from other contaminating crops. The isolation of a seed crop is usually done by providing distance between seed fields and contaminating fields.
- Types of isolation
    - Physical or distance isolation - expressed is in meters.
    - Time isolation - taking up of sowing in different dates (*i.e*) 30 days for most of the crops except in crops having indeterminate growth habit.
    - Barrier isolation - if the above said isolation is not possible we can go for barrier isolation by erecting tall shelter trees.
    - Physiological isolation - differ in flowering due to change in altitude.

Pre-sowing treatment

- Seeds should be appropriately treated.

Seed rate and sowing

- Seed rate should be based on seed lot viability and vigour. If not there will be lot of gaps in the field.
- Line sowing is advisable for seed crop adopting correct spacing then only we can achieve required population producing equal opportunity to each plant to develop and mature which is not possible in broadcasted crop. The sowing of seed crops in rows helps in conducting effective plant protection measures, roguing operations and field inspections.
- The seed crops should invariably be sown or transplanted at an optimum age of seedling at their normal planting time.
- Small seeds like cucurbits, solanaeceae and malvaceae vegetables should usually be planted shallow but large seeds could be planted a little deeper. Seeds would emerge from greater depths in sandy soils than in clay soils, and also in warm soil as compared to cold. In dry soils, seeds should be planted slightly deeper so that they come in contact with moisture.

Weeding

- Either manually or by using chemical means.
- Once or twice manual weeding during early stage of crop growth will help in controlling weeds and will provide favourable conditions for root formation and development.

Irrigation

- Optimum and timely irrigation is must.
- Over irrigation can promote vegetative growth, lodging, nutrient imbalance while under irrigation delay flowering, stunted growth reduced filling and immature drying. Hence irrigation at critical stages like sowing, life irrigation, flowering and fruit, pod formation, seed filling and seed maturation is must.
- With holding irrigation at harvest promotes earlier and quicker ripening, irrigation can also be used for staggering and achieving synchronization in hybrid seed production.

Nutritional factors

- Lack of N delay the field emergence, lack of P delay flowering, K is responsible for filling and lusture the crops.
- Zn is essential for fertilization. Selenium is responsible for germination in onion. Cu deficiency leads to poor embryo growth. Manganese deficiency causes marsh spot in peas. Boron deficiency results in poor seed development and hollow heart in peas. Iron deficiency causes sterility, Molybdenum also contribute bleaching in peas.

Roguing

- Roguing in most of the field crops may be done at any of the following stages as per needs of the seed crop.
  a. Vegetative/pre-flowering stage
  b. Flowering stage
  c. Fruit/pod formation stage.
- The rouging at vegetative / pre-flowering stage in cross-pollinated crops is extremely important to avoid genetic contamination.
- Flowering stage is also equally important, perhaps even more important than at the vegetative stage. In hybrid crops, where male sterility is being

used, special care is required in the removal of pollen shedders. In many crops removal of ear heads infested by seed borne diseases.

- Maturity stage is also equally important in the removal of the various contaminants affecting the physical purity of the seed.
- In root and vegetative crops a roguing at harvest time for confirmation of fruit / tuber / root characteristics is necessary. Sometimes, inspection is also being attended before seed extraction to select true to type fruits and pods to ensure genetic purity.

Diseases and pest control

- The seed crop should be maintained without any pest and disease attack.
- The leaf curl virus and yellow vein mosaic infestation drastically reduce the yield and quality in tomato and bhendi, respectively. The fruit / pod bores infestation causes seed discolouration and result in poor quality in terms of viability and storability.
- Similarly bruchid attack in lab leads to very poor storability. There is the heavy damage in flowering and seed set in crucifer vegetables due to diamond back moth incidence.

Harvest

- The time of sowing should be adjusted in such a way that the maturation does not coincide with rainy season or at high humidity weather periods.
- Avoid delayed or premature harvest, should be harvested at physiological maturation stage.

3. Post harvest handling of seed crop

Harvesting and threshing

- In vegetables, the harvests are to be taken in different pickings.
- For getting higher yield, the first and last one or two pickings may be taken for vegetables. This may promote further growth of the plant to put for the more number of fruit/pods. In addition, the seed recovery and quality will be poor in the above pickings.
- From the harvested produce the seeds should be extracted by suitable extraction methods.

Drying and grading

- Direct exposure of the seed to sunlight may affect its quality therefore seed may be dried under diffused sunlight in a shed with opening on all sides.
- Artificial drying is done by blowing dry air at 70-85°F but never exceeding 110°F.
- Grading is done with various types of grader.

Seed treatment

- Seed treatment with the fungicides and insecticides to arrest the carryover of pathogens and insects with the seed or their fresh entry in to it.

Packaging, labelling and sealing

- It is necessary to use right type of containers with (seed moisture content at proper level) labelling and sealing in the prescribed manner.

Movement and storage

- Precautions should be taken to avoid seed deterioration while in transit and or in storage.
- Ideal storage for long term is dry and cool condition.
- Ideal condition can be maintained if the sum of RH and Temperature does not exceed 100.

4. Seed quality control factors

Seed moisture

- The seed moisture affects seed storability.
- Seeds with low moisture store longer and remain free from insect pests.

Germination per cent

- Seeds are sown to provide next generation crops.
- The combination of pure seed per cent and germination per cent is called as Pure Live Seed (PLS) having better viability.

Vigour

- It indicates the ability of seed to emerge in varying environments or micro-climate of fields where it is grown.
- It is generally believed, but not always true, that high germination percentage is associated with high seed vigour.

Storage life

- Seed moisture content is the most important factor influencing loss of viability during storage.
- Most of the vegetable seeds which are costly are packed in suitable moisture vapour proof attractive containers and are not or least affected in storage or in transit but the seeds of some large sized seeds *e.g.*, garden pea, beans etc. are packed in porous containers, hence the seed moisture content fluctuates with the change in relative humidity of the atmosphere.

Seed health

- Vegetable seed should be free from seed borne diseases and insects infestation.
- Insect infestation normally destroys the embryos thus making the seeds unfit for sowing.
- Similarly most of the virus and bacterial diseases are seed borne. They are not only contaminate the crop but also help in spreading the disease fast.
- Hence, seeds must be free from pest and disease and treated with pesticide/fungicide to prevent contamination and spread.

Mechanism of control

The generally accepted system of seed certification involves inspections, sampling, testing, also enforcement of minimum standards which constitute the mechanism of quality control in seed.

8

Physiological Maturation and Harvesting Techniques in Seed Crops

Harvesting is one of the agronomic management practices that require technical knowledge on maturation of the crops. This knowledge is much important in seed production than in commercial production.

Physiological maturation

- It is the stage of accumulation of maximum dry matter within the seeds. The moisture content of the seed at this stage will be in decreasing order (25-30%) and is expressed with maximum dry weigh of seed, germination and vigour potential. The physiological maturation is represented for individual seed and this maturation will not be the same for the population, due to differential flowering habit.
- The attainment of physiological maturation is represented by sigmoid curve of growth pattern. The physiological maturation can be represented both as duration and visible symptoms.

Problem with physiological maturation

- All fruits of a population will not come to physiological maturation stage at single phase which will be more common in plants within determinate flowering habit than with determinate type.
- But fixing the days and harvesting is difficult. Hence the physiological maturation is identified based on visual symptoms that occur in fruit and seeds. Caution in this visible symptom is sometime due to insect attack the visible symptom will be obtained with fruits earlier to physiological maturation.
- The solution for the problem is harvesting of crop at harvestable maturity.

Physiological maturity symptoms in different crops

Crops	Physiological maturity symptoms
Amaranthus	Yellowish browning of inflorescence
Onion	Seeds become black on ripening in silver coloured capsules. 10% heads exposed black seeds.
Carrot	Second and 3rd order head turn brown
Radish	Pods become brown and parchment like
Turnip	Plants turn to brown parchment colour
Coriander	Plants turn to light yellow or brown in colour
Peas	Pods become parchment like
Beans	Earliest pods dry & parchment like and remaining have turned yellow
Brinjal	Fruit turn to straw yellow colour
Tomato	Skin colour turn to red and the fruits are softened
Cucumber	Fruit become yellowish brown in colour, and stalk adjacent to the fruit withers for confirming actual seed maturity.
Watermelon	Tendrils wither on fruit bearing shoot. Skin colour of the fruit resting on the soil is pale yellow and gives dull sound on thumping
Squash,Pumpkin	Rind becomes hard & its colour changes from green to yellow/ orange or golden yellow to straw colour
True potato seed (TPS)	Berries of potato becomes green to straw coloured and soft
Bitter gourd	Fruit pulp and seed becomes red and light brown respectively
Chillies	Green colour changes to red or yellow
Bottle and Sponge gourd	Rind becomes hard and colour changes to light brown or yellow
Colacasia	Drying and dieing of petiole and leaves
Zinger	Drying and falling down of pseudo-stem turning brown
Turmeric	Drying and falling down of stem turning brown
Garlic	The stem get dry and change in colour from green to brown
Seed potato	Haulms get dry, droop down turn dark brown in colour

Harvestable Maturation (HM)

- This harvestable maturation is for the population, next stage of physiological maturation.
- At this stage, more than 80% of the population will attain physiological maturation and hence without economic loss crop can be harvested.
- The seed quality will be more or less similar to physiological maturation. But moisture content will be lesser than PM (18-20%).

Caution on harvesting seed crop

- Harvesting should be done after PM at HM.
- The influence of pest attack and environmental influence should not be mistaken for PM or HM.
- The crop should never come to harvest (HM / PM) at rainy or high RH situation. This will reduce the seed quality characters drastically.

Advantages of correct method of harvesting

- Seed quality will be more and yield will be protected without loss due to shattering
- Processing loss will be reduced and seed storability will be more
- Above aspects will improve the seed marketability

Harvesting techniques

Harvesting of crop can be done either mechanically or manually.

Mechanical harvesting

- Machines/ harvesters are used for harvesting the crop
- Mechanical harvesting can be done only as single or once over harvest.
- Care should be taken at mechanical harvesting to avoid mechanical injury by adjusting blade size, gap between blades, speed of operation etc.

Manual harvesting

- Manual harvesting is done in two methods which is single harvest or periodical harvest.
- Either single or periodical harvest termed as picking is depending upon the growth habit of the crop.
- In determinate type where flowering is confined to a shorter duration once over or single harvest is recommended.
- In crops with indeterminate growth habit, periodical harvest or picking is recommended.
- Picking is harvesting the crop as and when a part of the population attains HM.

- The number of picking varies with crop depending on the growth habit of plant.

Usually seeds of later picking are not considered for seed purpose.

9

Threshing / Seed Extraction and Drying Methods

Removal of seeds from dry fruits is known as threshing while that of wet fruits is known as extraction. Threshing involves beating or rubbing the plant material to detach the seed from its pod or fruit. The detached seed is then winnowed to remove chaff, straw and other light material from the seed.

Threshing / extraction methods

1. Mechanical threshing

- Various types of threshing machines with adjustable cylinder speeds are available for extraction of vegetable seeds.
- Care must be taken to avoid damage to the seed during mechanical threshing, by properly adjusting the speed of the beaters, the width of the gap between the beaters and the concave, the airflow and the sieve sizes.

2. Hand threshing

Seed extraction from dry fruits

Common method mostly performed by women labour. Relatively cheap, easy and make use of surplus local labour. Hand threshing may be done in the following ways.

a. Rubbing – Rubbing seeds materials with a pressure in an open-ended trough line with ribbed rubber (bamboo contained). This method is quite suitable for pod materials such as *Brassica* and radish.

b. Beating – the seed materials is beaten with the help of wooden pliable sticks repeatedly with a tolerable force as the seeds are separated but not broken.

c. Flailing – specially designed instruments are used for separating the seeds from the plants. *e.g.*, Sweet corn.

d.Rolling – seed materials is rolled on threshing floor or tarpaulin repeatedly and seeds are easily separated.

e. Walked on – the seed material is spread on the threshing floor and children or other persons are asked to walk on the seeds materials till the seeds are separated.

Seed extraction from wet or flashy fruits

The seeds extraction from wet / flash fruits can be done by the following methods.

1. Manual method 2. Fermentation method 3. Chemical method

1. Manual Method

(a) Maceration *e.g.*, watermelon, (b) Crushing *e.g.*, brinjal, (c) Scraping *e.g.*, cucumber (d) Separated *e.g.*, muskmelon, (e) Scooping *e.g.*, pumpkins and (f) Extraction *e.g.*, squashes.

2. Fermentation Method

Fruits with pulp and seed are squeezed and kept as such for 24-48 hours. The seeds will settle down. Decayed pulp and immatured seed will float. The settled seeds are washed with more of water. The seeds are shade dried and then sun dried before using. Care should be taken to avoid germination of seed during fermentation. The seeds will be dull in colour.

3. Chemical method

i. Alkali method

- This method is relatively safe and can be used for small quantities of seed in cooler temperate areas where the fermentation method is not used.
- The pulp containing the extracted seed is mixed with an equal volume of a 10% solution of sodium carbonate (washing soda). The mixture is left for up to 48 hours at room temperature and after washed out in a sieve and subsequently dried.
- This method is not suitable for commercial seed production as sodium carbonate tends to darken the testa of the seed.

ii. Acid method

- Acid method is often favoured by large commercial seed producers as it produces a very bright clean seed.
- Addition of 30ml of hydrochloric acid per litre of seed and pulp mixture stirred properly and left for half an hour then the seeds are washed thoroughly with water, sieved and dried.
- The benefits of this method are (i) seed extraction and drying is done on the same day, (ii) higher seed recovery, (iii) the problems of low and high temperatures are avoided, (iv) discoloured seed resulting from fermentation is entirely avoided and (v) remove external seed borne pathogens.

Seed Drying

Removal or elimination of moisture from the seed to the required level is called drying. Drying of seeds is done by following methods

1. Sun drying (Natural Drying)
2. Forced air drying (Mechanical drying)
3. Use of desiccants (Chemical)

1. Sun drying (or) Natural Drying

Seeds are uniformly spread over clean dried yard and allowed for drying to the required moisture level. The seeds should not be dried under hot sun during 12.00 noon to 2.00 pm as it causes damage to seeds by UV rays. This method depends on weather conditions, which are unpredictable one.

Advantages

- Easy and cheap method
- Requires no additional equipment
- Does not require any expenditure on electricity or fuel

Disadvantages

- More chance for mechanical admixture
- Seed loss is more while drying due to insects, birds and animals
- Takes long time for drying

- Uneven drying
- High weather risk and damage due to sudden rain or heavy wind

2. Mechanical drying (or) artificial drying

Forced air is used for seed drying by the following three means.

a. Natural air drying: Natural air is blown upon the seeds using suitable air blower for drying. Continuous drying is possible in this method. In modern seed godowns provisions are made to forcible circulation of air with the help of electric blower or fan.If the outer air is comparatively dry, this method is followed. So it is possible only during dry months.

b. Drying with supplemental heat: Small quantity of heat is applied to raise the air temperature to 10-20^0 F for reducing the relative humidity of air used for drying. In this, drying is performed quickly due to use of dry air, but continuous drying for long period affects seed quality.

c. Heated air Drying: The air is heated considerably as much as by 100^0 F (40^0 C) and used for drying the seeds. Very quickly the seeds get dried. The seeds should not be continuously dried as it causes damage to seed. High moisture seeds should be dried by this method.

Advantages

- Quick and perfect drying is possible even under unfavourable weather condition
- Seed loss is minimum

Disadvantages

- Requires specialized equipment and machine, which is costly
- Care should be taken while drying using hot air, as it causes damage to the seed
- Tempering is to be followed while drying the seed in this method

Types of drier

a. Metal bin drier
b. Vegetable seed drier
c. Batch drier

a. Metal bin drier

- Seeds are placed in a metal bin and the heated air is blown in to the bin through the perforations made at the bottom of the bin.
- In this uniform drying of all layer is not possible for which decide the thickness of the seed layer to be taken to the bin and also have to stir the seed manually or mechanically at regular intervals.

b. Vegetable seed drier

- Seeds are separated over the bottom screen seed trays which are kept inside chamber or cabin.
- The heated air is passed to dry the seed. The heat is generated by electrical source and the air is passed through trays.
- Here uniform drying is possible.

c. Batch drier

- In *Bin batch dryers*, the seed is placed in a (usually round) bin, and ambient or slightly heated air is blown through it by a fan.
- The maximum thickness of the seed layer in the bin depends on the initial moisture content, the type of seed, the air temperature and RH and fan horse power.
- To obtain a uniform airflow through the seeds, a full perforated floor is required.
- Proper mixing of the seeds is essential before further storage or packaging. This can be addressed by installing one or more grain stirrers to mix the entire content of a bin for 3-12 hours.
- A seed transport wagon can be transformed into a *Wagon batch dryer* by equipping it with a plenum, a perforated floor, and a fan/heater unit coupled with a canvas transition to the wagon.
- Wagon batch dryers are most frequently used for drying fragile seeds such as large-seeded legumes (*e.g.*, field or garden beans and peanuts). The recommended air flow rate for the ambient- air wagon drying of a 1.5m layer of peanut seeds is $0.25m^3$ of air per m^2 of floor area.

3. Use of seed desiccants (Chemical drying)

In this method silica gel or fused calcium chloride ($CaCl_2$) is used to absorb the moisture from the seed and its surrounding environment. Silica gel is of two types, as

i) Indicator type

ii) Non-indicator type

- Active ingredient in Silica gel is Lithium chloride, which is responsible for drying process. Silica gel can absorb moisture upto 15 per cent of its weight. So to get very low moisture content we can use this, which is not possible in mechanical driers.
- Indicator type will be blue in colour and on absorbing moisture, this turns to pink colour. So we can remove this and reuse after dehydration.
- Non- indicator type will be white in colour and remains same (white) even after absorption of moisture content. So there is no indication in this type. But this can also be reused after dehydration. Calcium chloride is used for most of the vegetable and flower seeds of breeding material. Here the quantity needed is more. It can absorb 10% of its weight. The method is suitable for drying small quantities of seeds only. It is a sophisticated and costlier method.

Advantages

- Less time consuming
- Drying rate is uniform

Disadvantages

- It cannot be used in large scale
- A skilled person is required to monitor the operation

The rate of drying depends upon the following factors :

- The moisture content of the seed
- The existing relative humidity and temperature of the environment
- Kind of seeds, Depth of spread of seeds, Rate of air blow, Drying temperature
- Size and capacity of the drier

Tempering

When the heated air is used for drying, moisture content in the surface layer of the seed is removed at a faster rate, while the moisture present inside tends to reach outside slowly to maintain the equilibrium. On continuous drying a pressure gradient is developed inside the seed due to difference in moisture content between the dried outer layer and drying inner layer of the seed. This results in the damage of seeds by formation of hair line cracks in the seed. Hence tempering is to be followed. It refers to the discontinuation of drying operation for a specified period to allow the moisture present in the interior of the seed to migrate all over the exterior portion uniformly.

10

Seed Processing and Seed Treatment

Seed Processing

Seed processing may be understood to comprise all the operations after harvest that aim at maximizing seed viability, vigour and health. It includes cleaning, drying, seed treatment, packaging and storage.

Purposes of seed processing

- To improve the quality and uniformity of the seed by cleaning and grading.
- To reduce the cost of transport. This is achieved by reducing the bulk of the seed lot by cleaning debris and by removing empty or fractured seed.
- To increase the longevity of seeds by drying seeds to safe moisture level and treating with protective chemicals
- To reduce the variability in vigour by invigourating the seeds and removing the low vigour seeds
- The reduce the heterogeneity of the seed lot.

Principles and objectives

The quality of seed is improved during processing in two ways (1) separation of other crop seeds or inert matter and (2) upgrading or the elimination of poor quality seeds.

Causes for heterogeneity

- Variability in soil for fertility, physical, chemical and biological properties
- Variability in management practices (irrigation, application of nutrients etc.)
- Variability in ability of the seedling for utilizing the inputs

- Variability in pest and disease infestation
- Position of pod or fruit in a plant or the position of seed in a pod.

This heterogeneity can be narrowed down in the processing of seeds by eliminating the undersized, shrivelled, immature, ill filled seeds using appropriate sieve size. The germinability and vigour of the seed lot can be upgraded by grading the seeds according to size, specific gravity, length and density of the seeds.

Requirement in seed processing

- There should be complete separation.
- There should be minimum seed loss.
- Upgrading should be possible for any particular quality.
- There should be having more efficiency.
- It should have only minimum requirement.

Sequence of operation in seed processing

- Drying, receiving, pre-cleaning, conditioning, cleaning, separating or upgrading, treating (Drying), weighting, bagging and storage or shipping.

In seed cleaning the seed is separated from undesirable material (*i.e.*, inert matter, weed seeds, other crop seeds, light and chaffy seeds, deteriorated and broken seeds) on the basis of physical differences like density, surface texture, affinity to liquids and electric conductivity is known as seed cleaning. Physical differences like length, width, shape, weight etc are common in crop species and they form the basis for seed cleaning operations. Cleaning of seeds involves three steps.

- Pre-cleaning and pre-conditioning
- Basic cleaning
- Seed upgrading

1. Pre-cleaning & pre-conditioning

It refers to the operations such as shelling, debearding etc. that prepares the seed lots for basic cleaning and also for the removal of particles such as trash, stones, clods etc larger than crop seed. Some pre-cleaners also remove particles that are lighter in weight and smaller in size than the crop seed. Precleaning is not required for hand harvested and winnowed seed lots. Equipment used for pre-cleaning and pre-conditioning are

- Scalper or Rough Cleaner- Scalpers are used to remove large trash.
- Huller/ Scarifier- Huller is a device to remove husk or outer seed coat. Scarifier scratches the seed coat.
- Debearder- The debearder removes the hair like structures present on the seeds.

2. Basic Seed Cleaning

It refers to actual cleaning and grading of seeds and is essential process in seed cleaning operation. The basic seed cleaning is done over an air screen machine commonly referred to as an air screen cleaner. It is the basic equipment in all seed processing plants. The separation of undesirable material from seed is done on the basis of differences in seed size and weight. The air screen machine uses three cleaning elements:

1. Aspiration: the light seed and chaffy material is removed from the seed mass through aspiration.
2. Scalping: Good seed are dropped through screen openings but large material (trash, clods etc.) are scalped off over the screen into a separate spout.
3. Grading: The good seed ride over the screen openings, while smaller particles (undersized, weed seeds, shriveled) drop through the screen perforations.

Principle of operation of air screen machine

- The air blast removes lightweight seed and chaffy seed.
- Scalping screen removes material larger than the crop seed.
- Grading screen drop out material smaller than crop seed.
- Eccentrics do the shaking motions of the screens.
- The two shoes in 4 screen cleaner move in opposite direction to balance each other also to reduce machine vibrations to minimum.
- In four screen cleaner, the screens do the following, First screen does scalping, Second screen does grading, Third screen does close scalping, Fourth screen does close grading.

3. Seed upgrading

In certain instances it is necessary to remove specific contaminants by precise size grading. The various processing operations conducted after basic cleaning to further improve seed quality are regarded as upgrading operations. The choice of upgrading operations however shall depend upon the type of contaminants and crop seed. The various types of upgrading operation equipment, their principle of operation and specific uses are given below :

1. Size : Based on size it can be separated with air screen cleaner cum grader
2. Length : Disc or indented cylinder separator
3. Weight : Specific gravity separator
4. Shape : Spiral separator or draper separator for round and flat seeds
5. Surface texture : Rough from smooth surface seed- dodder mill
6. Colour : Electronic colour separator
7. Electrical conductivity: Seed differing in their ability to conduct electrical charge can be separated with electronic separator.
8. Affinity to liquid: The seed coat of seed will absorb water, oils etc., which provides a means of separating seed on the magnetic separator.

Type of upgrading Operations & types of Machines	Principle of operation of the equipment
1. Sizing and grading	
(a) Width and thickness sizing and grading i. Horizontal flat screen separator ii. Vertical ribbed screen separator iii. Cylindrical screen separator.	These machines make extremely sensitive separations on the basis of differences in width and thickness. The separators employ gravity, centrifugal force, product pressure or a combination of these forces to make the separation.
b. Length sizing and grading i. Disc Separator ii. Cylinder separator	Seeds are separated on pure length basis. In disc separator the disc lift uniformly shaped and sized and under sized particles out of a mass of seed. The cylinder separator operates on a centrifugal force principle in which the speed of the cylinder holds seed in the indents, lifting them out of the mass until the indents are inverted to the point where gravity causes the particles to fall.
2. Gravity or weight separations i. Gravity separator	The gravity separator employs a flotation principle. In this separation seeds are vertically stratified in layers on the deck according to density. Seeds of same size are stratified and separated by differences in their specific gravity. The oscillating movement of the table walks the heavy seeds in contact with the deck uphill while the air floats the light seeds downhill. The seeds travelling to the edge

of the table range from light at the lower end to heavy at the upper end. The discharge can be divided into any number of density fractions.

3. Air separations

i. Pneumatic separator

ii. Aspirator

iii. Fractional aspirator

It uses the movement of air to divide seeds according to their terminal velocities. Terminal velocity refers to the velocity of air required to suspend particles in a rising air current. When the seed is introduced into an air stream all the particles with lightweight are lifted by the velocity of the air, while the seeds with heavy weight fall below. In pneumatic air separator the air is blown through the fan which lifts the light material and the seeds which are light in weight. In aspirator the fan is at discharge end and induces a vacuum, which allows the atmospheric pressure to force air through the separator. In fractional aspirator when a seed mixture is introduced into the lower end of an expanding air column, heavy seeds fall against the airflow and lightseeds are lifted. Air velocity through the expanding columns lessens and gradually drops out seeds with lower terminal velocities. Each outlet along the column receives a lighter fraction of seeds the mixture is thereby separated into several grades.

4. Surface Texture Separation

i. Roll Mill

ii. Magnetic separator

iii. Inclined Draper

The velvet roll mill classifies seeds according to a difference in texture of seed coat. When a seed mixture is fed on to the upper end of the rolls, the smooth seeds travel downhill between them and are discharged at the lower end. Rough-coated seeds caught in the velvet take a bouncing path between shield and rolls and are thrown over sides. The discharge from the sides of the rolls is caught is several directions. The roughest seeds are ejected first. Magnetic separator take advantage of the surface texture and stickiness of seed to make a separation. A seed mixture and proportionate amount of water and finely ground iron powder are mixed in a screw conveyor or other mixing device. In the presence of moisture, the iron powder will adhere to rough cracked and sticky seeds. When the mixture is fed to the top of a horizontal revolving magnetic drum smooth or sticky seeds that are relatively free of iron powder all from the drum, while the rough textured seeds with iron powder stick to the drum until they are removed by a rotary brush. The inclined draper separator senses differences in shape and surface texture to separate seed on an inclined plane. A mixture to be separated is metered on to the center of an inclined draper belt travelling in an uphill direction. Round or smooth seeds, which roll or slide down the draper, faster than the draper travelling upwards. Flat or rough-coated seeds are carried to the top of the inclined plane and dropped into a separate hopper. Separation of smooth or round seeds from rough, flat or elongated seed.

5. **Electronic separation**
Electric colour sorters

The electric colour sorter separates the seeds on the basis of difference in colour brightness. One type of machine picks up the seeds on a series of suction fingers and carries them to a phototube where they are judged for colour brightness and ejected into separate containers one at a time.

6. Other Separators
1. Spiral separator
2. Polishers

The spiral separator makes a division of seed according to shape or the degree of its ability to roll. The separator resembles a stationary open screw conveyor standing on end. The mixture fed onto the spiral at the top, slides or rolls down the inclined surface. The fast rolling seeds gain speed and are thrown by centrifugal force into an outer housing, which directs them to a chute below. The slow rolling seeds remain on the inner inclined surface and enter a second chute at the bottom. Polishers use a polishing agent such as sawdust or bran to remove discoloration. In some of the polishers mild mechanical rubbing action is provided. To improve the lusture of the seeds.

The Sequence flow of seed lot in a processing plant is as below :

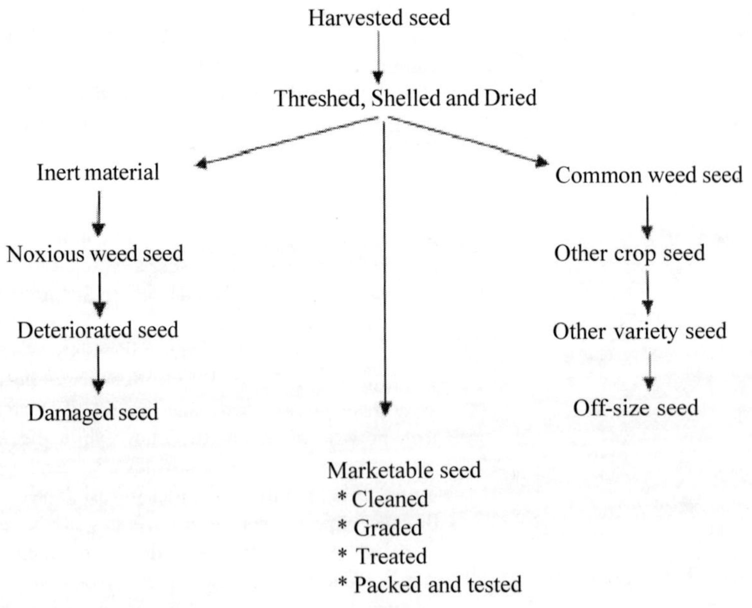

Seed Treatment

Seed treatment refers to application of any treatment by chemical, physical or biological means to alter the germination or storage behaviour of seeds.

Benefits or advantages of seed treatment

- Provides protection from pest
- Protects the seeds from pathogens
- Improves germination
- Addition of nutrition
- Facilitate easy sowing

Types of seed treatment

1. Presowing seed treatment

High quality seed is the key for successful production. Uniformity of growth and synchrony in development are highly desirable characters that promise better establishment. To increase the productivity by obtaining good seedling establishment, some presowing seed treatments are administered. Seed treatments are also imperative to overcome the seed dormancy problems,

- Dormancy breaking treatments – Scarification, Stratification
- Germination augmenting treatments – Hardening, priming, fortification
- Seed coating treatments – Pelleting, colouring
- Seed protection treatments – Fungicide, insecticide treatment

2. Prestorage seed treatments

Prestorage treatments of harvest-fresh seed are primarily aimed towards protection against deteriorate senescence during storage. Seed storage which is again threatened by insect and pathogen attack, can also be taken care of by prescribed prestorage seed treatments.

- Halogenation–It is the protective treatment given to seed in this treatment, chlorine or iodine based halogen mixtures are prepared and treated with seed either dry or as slurry treatment.
- Antioxidant treatment.
- Seed sanitation–Seed disinfection, Seed disinfestations, Seed protectants

- Seed fumigation: In cases of large quantities available for seed treatment, gaseous treatment is recommended. Fumigation is the seed treatment given in the form of gas under air tight chamber. They have the advantage of (i) very rapid action against insects and (ii) it has high penetrability. The commonly used fumigants are Celphos, Quickphos, Ethylene dibromide.

3. Mid storage seed treatments

Seeds in storage accumulate peroxidative damage to cell membranes during senescence. Mid storage seed treatments are capable of ameliorating the age induced damages and restoring the seed vigour to a certain extent besides, the seed viability and productivity of stored seeds are also improved. Eg. Hydration-dehydration treatment.

Equipments used for seed treatment

- Slurry Treaters
- Direct Treaters
- Home-made drum mixer

Precautions in seed treatment

- Care must also be taken to treat seed at the correct dosage rate; applying too much or too little material can be as damaging as never treating at all.
- Seed with very high moisture content is very susceptible to injury when treated with some of the concentrated liquid products.
- Most products used in the treatment of seeds are harmful to humans, but they can also be harmful to seeds.
- Extreme care is required to ensure that treated seed is never used as human or animal food. To minimise this possibility, treated seed should be clearly labelled as being dangerous, if consumed.

Special seed treatments

Seed hardening

It is a method followed for improving the germination of seeds sown in arid regions. Pre-sowing hardening of seed helps in modifying the physiological and biochemical nature of seed protoplasmic characters and increasing the physiological activity of the embryo and associated structures. This leads to increase in elasticity of cell wall and developments of a stronger and efficient root system.

Seed fortification

Seeds are invigorated using vitamins, minerals and chemicals, *e.g.*, Tomato seeds are soaked in Vitamin E dissolved in acetone during storage.

Seed infusion

Seeds with internal seed borne diseases which need penetration of chemical to kill the pathogen are soaked in alcohol dissolved pesticides. Some crop seeds are susceptible to soaking injury that case seeds are soaked in alcohol dissolved nutrients like vitamin and can be dried quickly, *e.g.*, Infusion of Topsin, Thiram into vegetable seeds.

Seed Colouring

Seeds can be coloured with non-toxic dyes like methyl red, bromocresol green, methylene blue and rosebengal dyes. They can help in lot identity, variety identity, brand specificity, identity of A,B&R lines, to indicate the efficiency of seed treatment etc.

Seed Tapes

These are tapes embedded in paper or gel. Pretreated seeds are placed in equal distance in the gel and cooled. These tapes can be rolled out in the field, covered with soil and watered.

Seed treatment with bio-fertilizers

To facilitate fixing of atmosphere nitrogen and conserve fertilizer utilization this treatment is done, *e.g.*, Rhizobium, Azotobacter, etc.

Seed coating

Seed coating is a technique by which additives such as pesticides nutrients or nitrifying bacteria are applied to the external surface *i.e.*, seed coat.

11

Seed Packing and Storage

Seed Packing

Seed packaging is the process of filling, weighing and sewing of bags with seed. The choice of packaging materials and amount of seeds to be packed depends on kind of seeds to be packed, duration of storage, storage environment, the seed moisture content, the cost of seed, the cost of packaging material, the geographical area where the seeds will be stored.

Types of Packaging Material

Moisture vapour permeable container

Allows transmission of moisture from seed to atmosphere and atmosphere to seed either in the form of liquid or gaseous form. *e.g.*, jute (burlap) bag, cloth bag paper bag and multiwall paper bag.

Moisture vapour resistant container

Allows transmission of moisture on either side only in gaseous form *e.g.*, jute bag laminated with thin polythene film

Moisture vapour proof container

The container allows moisture neither in the form of liquid in the gaseous stage *e.g.*, tin can, polythene bags, aluminium foil pouches, glass bottles.

The packaging materials should protect most physical qualities of seed and should have sufficient tensile strength, bursting strength and tearing resistance to withstand the handling stresses. Such materials may not always protect the seeds against either insect pests or moisture regain.

Seed storage

Seed storage is preservation of seed with initial quality until it is needed for planting.

Stages of seed storage

The seeds are considered to be in storage from the moment they reach physiological maturity until they germinate or until they are thrown away because they are dead or otherwise worthless. The entire storage period can be conveniently divided into following stages.

- Storage on plants (physiological maturity until harvest)
- Harvest, until processed and stored in a warehouse.
- In - storage (warehouses)
- In transit (Railway wagons, trucks, carts, railway sheds etc.)
- In retail stores
- On the user's farm

General principles of seed storage

- Seed storage conditions should be dry and cool
- Effective control of storage pests
- Proper sanitation in seed stores
- Before placing seeds into storage they should be dried to safe moisture limits, appropriate for storage system.
- Store only high quality seed *i.e.*, seeds which are well cleaned, treated, with high germination and vigour.
- Determine seed storage needs in view of period or length of storage time and prevailing climate of the area during storage period.

Long-term storage requires more exacting conditions of seed storage than short-term storage. Similarly, the regions with favourable storage climate, *i.e.*, one where relative humidity is rather low, require less sophistication than areas of high relative humidity.

Types of storage

1. Storage at ambient temperature and humidity

Seeds can be stored in piles, single layers, sacks or open containers, under shelter against rain, well ventilated and protected from rodents and store at least for several months.

2. Dry storage with control of moisture content but not temperature

Orthodox seeds will retain viability longer, when dried to low moisture content (48%) and then stored in a sealed container or in a room in which humidity is controlled, than when stored in equilibrium with ambient air humidity. Cool condition is especially favourable.

3. Dry storage with control of both moisture content and temperature

This is recommended for many orthodox species which have periodicity of seeding but which are planted annually in large scale afforestation projects. A combination of 4-8% moisture content and 0 to 5° temperature will maintain viability for 5 years or more.

4. Dry storage for long-term gene conservation

Long-term conservation of gene resources of orthodox agricultural seeds is -18°C temperature and 5±11% moisture content

5. Moist storage without control of moisture content of temperature

Suitable for storage of recalcitrant seeds, for a few months over winter. Seeds may be stored in heaps on the ground, in shallow pits, in well drained soils or in layers in well ventilated sheds, often covered or mixed with leaves, moist sand, peat or other porous materials. The aim is to maintain moist and cool conditions, with good aeration to avoid overheating which may result from the relatively high rates of respiration associated with moist storage. This may be accomplished by regular turning of the heaps.

6. Moist cold storage, with control of temperature

This method implies controlled low temperature just above freezing or less commonly, just below freezing. Moisture can be controlled within approximate limits by adding moist media *e.g.*, sand, peat or a mixture of both to the seed, in proportions of one part media to 1 part seed by volume, and re-moistening periodically or more accurately by controlling the relative humidity of the store. This method is much applicable to temperate recalcitrant genera.

7. Cryopreservation

It is also called as cryogenic storage. Seeds are placed in liquid nitrogen at -196°C. Seeds are actually placed into the gaseous phase of the liquid nitrogen -150°C for easy handling and safety. Metabolic reactions come to a virtual standstill at the temperature of liquid nitrogen and the cells will remain in an unaltered state until the tissues are removed from the liquid nitrogen and defrosted.

Therefore, little detrimental physiological activity takes place at these temperatures, which prolongs the storage life of seeds. It is not practical for commercial seed storage, but is useful to store the valuable germplasm.

Maintenance of viability in storage

- Store well mature seeds
- Store normal coloured seeds
- Seeds should be free from mechanical injury
- Seeds should be free from storage fungi or micro organisms. Seeds should be treated with fungicides before storage
- Seeds should not met with adverse conditions during maturation
- Storage godown should be fumigated to control storage insects, periodically
- Storage environment or godown should be dry and cool
- Seeds should be dried to optimum moisture content
- Required relative humidity and temperature should be maintained during storage
- Suitable packaging materials should be used for packing.

Factors influencing seed storage

1. Biotic factors

- Factors related to seed- Genetic makeup of seed, Initial seed quality, Provenance, Seed moisture content
- Other biotic- Insects, Fungi, Rodents, Mishandling during sampling, testing

2. Abiotic factors

- Temperature, Relative humidity
- Seed store sanitation, Gaseous atmosphere
- Packaging material, Seed treatment

Biotic factors- Seed factors

1. Genetic factors

The storage is influenced by the genetic makeup of the seed. Some kinds are naturally short lived *e.g.*, Onion, Soybeans, Ground nut etc., Based on the genetic makeup seeds are classified into

- Micro biotic - short lived
- Meso biotic - medium lived
- Macro biotic - long lived

2. Initial seed quality

Seeds of high initial viability are much more resistant to unfavourable storage environmental conditions than low viable seed. Once seed start to deteriorate it proceeds rapidly. The seed which injured mechanically suffered a lot and loses its viability and vigour very quickly.

3. Effect of provenance

The place where the seed crop was produced greatly influences the storability. This is due to different climatic conditions and soil types prevailing in different places.

4. Effect of weather

Fluctuating temperature during seed formation and maturity will affect seed storage. Pre-harvest rain may also affect the viability.

5. Seed moisture content

The amount of moisture in the seeds is the most important factor influencing seed viability during storage. Generally if the seed moisture content increases storage life decreases. If seeds are kept at high moisture content the losses could be very rapid due to mould growth, very low moisture content below 4% may also damage seeds due to extreme desiccation or cause hard seededness in some crops. Since the life of a seed largely revolves around its moisture content it is necessary to dry seeds to safe moisture contents.

The safe moisture content however depends upon storage length, type of storage structure, kind / variety of seed type of packing material used. For cereals in ordinary storage conditions for 12-18 months, seed drying up to 10% moisture content appears quite satisfactory. However, for storage in sealed containers drying upto 5-8% moisture content depending upon particular kind may be necessary.

Other biotic factors - Microflora, Insects and Mites

The activity of all these organisms can lead to damage resulting in loss of viability. The microflora activity is controlled by relative humidity, temperature and moisture content of seed. Treated seeds with fungicides can be stored for longer periods.

Fumigation to control insects will also help to store longer period. Fumigants - *e.g.*, Methyl bromide, Hydrogen cyanide, Ethylene dichloride, Carbon tetra chloride, Carbon disulphide and Naphthalene and Aluminium phosphine.

Abiotic factors

1. Relative humidity

Relative humidity is the amount of H_2O present in the air at a given temperature in proportion to its maximum water holding capacity. Seeds attain specific and characteristic moisture content when subjected to given levels of atmospheric humidity. This characteristic moisture content called equilibrium moisture content. Thus the maintenance of seed moisture content during storage is a function of relative humidity and to a lesser extent of temperature. At equilibrium moisture content there is no net gain or loss in seed moisture content.

2. Temperature

Temperature also plays an important role in life of seed. Insects and moulds increase as temperature increases. The higher the moisture content of the seeds the more they are adversely affected by temperature. Decreasing temperature and seed moisture is an effective means of maintaining seed quality in storage.

3. Gas during storage

Increase in O_2 pressure decrease the period of viability. N_2 and CO_2 atmosphere will increase the storage life of seeds.

An ideal storage facility should satisfy the following requirements

- It should provide maximum possible protection from ground moisture, rain, insect pests, moulds, rodents, birds, etc., It should provide the necessary facility for inspection, disinfection, loading, unloading, cleaning and reconditioning.
- It should protect grain from excessive moisture and temperature favourable to both insect and mould development, it should be economical and suitable for a particular situation.

12

Seed Certification

Seed certification

It is a legally sanctioned system for quality control and seed multiplication and production. It involves field inspection, pre and post control tests and seed quality tests. As per Indian Seed Act seed certification is voluntary and it is not compulsory. The seed that is sold in the market is of two types certified seed or truthfully labelled seed. Whereas truthfully labelled seed is one which is being produced and marketed by the producing company by maintaining the labelling standards.

Objective of seed certification

To maintain and make available to the farmers, high quality seeds and propagating materials of notified kind and varieties. The seeds are so grown as to ensure genetic identity and genetic purity.

Eligibility for certification of crop varieties

Seed of only those varieties which are notified under section 5 of the seeds Act, 1966 shall be eligible for certification. Breeder seed is exempted from certification. Foundation and certified class seeds come under certification.

Concept of seed certification

Concept of seed certification was originated in Sweden during twentieth century by visiting agronomist and plant breeder to the progressive farmers, who took seeds from them, primarily with the objective of educating them on how to avoid contamination. This initiated field inspection process.

Principles of seed ceritification agency

1. It should not involve in seed production and marketing
2. It should have autonomy

3. Seed certification procedure adopted should be uniform throughout the country
4. It should closely associated with technical institutes
5. It should operate on a no profit and no loss basis
6. It should have adequate technical staff and facilities for timely inspection of seed fields
7. It should serve the interests of seed producers and buyers

Procedure/ Phases of seed certification

1. Receipt and scrutiny of application
2. Verification of seed source
3. Field inspection
4. Post harvest supervision of seed crops
5. Seed sampling and testing
6. Labelling, tagging, sealing and grant of certificate.

1. Receipt and scrutiny of application

a. Application for registration

Any person, who wants to produce certified seed shall register his name with the concerned Assistant Director of Seed Certification by remitting Rs.25/- per crop, per season. There are 3 seasons under certification *viz.*, *kharif* (June - September), *rabi* (October - January) and summer (February - May). The applicant shall submit two copies of the application to the ADSC 10 days before the commencement of the season or at least at the time of registration of sowing report. On receipt of the application, the Assistant Director of Seed Certification will verify the time limit variety eligibility and its source, the class mentioned, remittance of fee etc., The application, if accepted will be given an application no (*e.g.*, Paddy / k/01-97-98 where paddy refers the crops to be registered, K the season, 01-the application Number and 99-2000 the financial year). The original application is retained and the duplicate is returned to the applicant.

b. Sowing report (Application for the registration of seed farm)

The seed producer who wants to produce certified seeds shall apply to the Assistant Director of Seed Certification in the prescribed sowing report form in quadruplicate with prescribed certification fees along with other documents such as tags to establish the seed source. Separate sowing reports are required for different crop varieties, different classes, different stages and if the seed farm fields are separated by more than 50 meters. Separate sowing reports are also required if sowing or planting dates differ by more than 7 days and if the seed farm area exceeds 25 acres. The sowing report shall reach concerned Assistant Director of Agriculture Seed Certification within 35 days from the date of sowing or 15 days before flowering whichever is earlier. In the case of transplanted crops the sowing report shall be sent 15 days before flowering. The producer shall clearly indicate on the reverse of sowing report, the exact location of the seed farm in a rough sketch with direction, distances marked from a permanent mark like mile stone, building bridge, road, name of the farm if any, crops grown on all four sides of the seed farm etc., to facilitate easy identification of the seed farm by the seed certification officer.

The Assistant Director, Seed Certification on receipt of the sowing report, scrutinises and register the seed farm by giving a Seed Certification number for each sowing report. Then he will send one copy of the sowing report to the Seed Certification officer, on to the Deputy Director of Seed Certification and the third to the producer after retaining the fourth copy.

2. Verification of seed source

During his first inspection of seed farm the Seed Certification officer will verify whether the seed used to raise the seed crop is from an approved source.

3. Field inspection

- The objective in conducting field inspection is to verify the factors which can cause irreversible damage to the genetic purity or seed health.
- The seed certification officer authorized by the registering authority shall attend the field inspections.

Crop stages for inspection

The stages and number of field inspections required depends on the breeding system of the seed crop.

Sexually propagated crops

Crop	Vegetative	Flowering	Post flowering	Pre harvest
Self pollinated	-	3	-	3
Cross pollinated	3	3	-	3
Hybrid	3	3	3	3

General factors to be observed

Vegetative	Seed source, cropping history, isolation distance, seed production practices
Flowering	Isolation distance, offtypes, rogues, objectionable weeds, seed borne diseases, other crop plants.
Post flowering and pre-harvest	Isolation distance, offtypes, rogues, objectionable weeds, seed borne diseases, other crop plants.
Harvest	Isolation distance, offtypes, rogues, objectionable weeds.

Vegetatively propagated crops

Potato	-	Sprouting, seedling lifting and replanting, tuberisation, tube hardening, haulm cutting stages
Cauliflower	-	Curd formation, bolting
Knolkhol	-	Knob formation, bolting
Cabbage	-	Head formation

Field counts

It is a representative sample of plants taken at random from a seed plot for recording the observation on off types, pollen shedders, diseased plants, inseparable other crop plants. As per provision of seed certification it is necessary to examine each and every plant in the seed plot for contaminant. It is however impracticable to do so. During each field inspection field counts are taken randomly covering of the seed plot and observations are made on the plants from each selected field counts.

Points to be observed before counting

1. All plants falling in each count must be examined for each factor
2. In hybrid seed field the prescribed number of the field counts should be taken in each parent separately.

Number of field counts

Area of the field (acres)	No. of counts to be taken
<5	5
5–10	6
10–15	7
15–20	8
20–25	9

Number of plants /heads per count

S.No.	Crop	No. plants per count
1.	Berseem, jute, lucerne, mesta, soyabean	1000 plants
2.	Beans, cluster beans, green leafy vegetables	500 plants
3.	Bhendi, brinjal, bulb crops, capsicum, chilli, cole crops, cucurbits, potato, root crops, tomato	100 plants

Sources of contamination or factors to be observed

Physical contaminants

Physical contaminants are inseparable other crop plants, objectionable weed plants and diseased plants.

- Inseparable crop plants - These are plants or different crops which have seed similar to seed crop.
- Objectionable weed plants - These are weeds, whose seeds are difficult to be separated once mixed, which are poisonous, which have smothering effect on the main crop, which are difficult to eradicate once established, difficult to separate the seeds. These seeds cause mechanical admixtures.
- Designated diseases -The diseases which may reduce the yield and quality of seeds are termed as designated diseases.

Genetic contaminants

Genetic contaminants consists off-types, volunteer plant, pollen shedders.

a. Off type

Plant that differs in morphological characters from the rest of the population of a crop variety. Off type may belong to same species or different species of a given variety. Plants of a different variety are also included under off types. Volunteer plants and mutants are also off types.

b. Volunteer plant

Volunteer plants are the plants of the same kind growing naturally from seed that remains in the fields from a previous crop.

c. Pollen shedders

In hybrid seed production involving male sterility, the plants of 'B' line present in 'A' line are called pollen shedders. Sometimes 'A' line tends to exhibit symptoms of fertile anthers in the ear heads of either on the main tiller or side tiller and these are called partials. These partials are also counted as pollen shedders.

Double count

In any inspection, if the first set of counts shows that the seed crop does not confirm to the prescribed standard for any factor, a second set of counts should be taken for the factor. However, when the first set of counts shows a factor more than twice the maximum permitted, it is not necessary to take a second count. On completion of double count assess the average for the two counts. It should not exceed the minimum permissible limit.

Inspection report

The seed certification officer after taking field counts and comparing them with the minimum field standards, the observations made on the seed farm field should be reported in the prescribed proforma to :

- Deputy Director of Agriculture (Seed Certification)
- To the Seed Producer
- Assistant Director of Agriculture (Seed Certification) and
- Fourth copy retained with see certificate officer

Assessment of seed crop yield

It is necessary to avoid malpractice's at the final stage during harvest operation. The seed certification officer is expected to fix the approximate seed yield.

Liable For Rejection Report (L.F.R.)

If the seed crop fails to meet with any one factor as per the standards, Liable for Rejection report is prepared and the signature of the producer is obtained and sent to Deputy Director of Agriculture Seed Certification within 24 hours.

Re-inspection

For the factors which can be removed without hampering the seed quality, the producer can apply for re-inspection to the concerned Deputy Director of Agriculture Seed Certification within 7 days from the date of first inspection order. For re-inspection half of the inspection charge is collected.

4. Post harvest supervision of seed crop

The post harvest inspection of a seed crop covers the operations carried out at the threshing floor, transport of the raw seed produce to the processing plant, pre cleaning, drying, cleaning, grading, seed treatment, bagging and post processing storage of the seed lot.

Pre-requisites for processing

1. Processing report should accompany the seed lot.
2. Seed should be processed only in approved processing unit.
3. It should be correlated with the estimated yield.
4. Field run seed should be brought to the processing unit within the 3 months from the date of final inspection. Processing and sampling should be done within 2 months in oil seed crops and 4 months for other crops from the date of receipt in the processing unit.

Intake of raw produce and lot identification

The seed certification officer in-charge of the seed processing plant may, after verification of the above stated documents and total amount of seed accept the produce for processing. After verification he should be issue a receipt to the seed grower. Each seed lot has to be allocated a separate lot number for identification.

Processing of seed lot

1. It is done to remove chaff, stones, stem pieces, leaf parts, soil particles etc from the raw seed lot.
2. Grading to bring out uniformity in the seed lot.
3. Seed treatment to protect it from storage pests and diseases.

Processing inspection

1. The processing should be done in the presence of concerned seed certification officer.
2. The recommended sieve size should be used for grading

5. Seed sampling and testing

During packaging Seed Certification officer will draw samples according to ISTA Procedure and send the sample to Assistant Director of Agriculture (See Certification) concerned within a day of sampling. The ADASC will in turn send the sample to the Seed Testing Laboratory within 3 days of receipt of the sample to testing seed standards *viz.*, physical purity, germination, moisture content and seed health as prescribed. The Seed Testing Officer will communicate the result to the ADASC concerned within 20 days. On receipt of the analytical report the ADASC will communicate the result to the producer and Seed Certification officer.

6. Labelling, tagging, sealing and grant of certificate

After receiving the seed analytical report the Seed Certification officer will get the tag from the ADASC and affixes labels (producer's label) and tags (blue for Certified Seed and White for Foundation Seeds) to the containers and seals them to prevent tampering and grants certificate fixing a validity period for 9 months. If the seed is not sold within the stipulated period, it can be revalidated for a period of six months if the seed lot meets the required seed standards. The seed can be revalidated as long as it meets the prescribed seed standards and for each revalidation the validity period will be extended for six months. Tagging should be done within 60 days of testing.

Resampling and reprocessing

When a seed lot does not meet the prescribed seed standards in initial test, on request of the producer Seed Certification Officer may take resample. If the difference in germination analysed and required is within 10, then straight away resampling can be done. If it is >10, reprocessing and resampling may be done.

The producer should request the Seed Certification Officer concerned in writing within 10 days from the receipt of the result. No charge is collected for resampling. When a seed lot, fails even after free sampling reprocessing can be taken upon special permission from Deputy Director Seed Certification. For such reprocessing a fee of Rs.20/- Q and lab charges Rs.10/- Q is collected.

Revocation of certificate

If the certification agency is satisfied that the certificate granted by it has been obtained by misrepresentation of essential facts, or the holder of the certificate has failed to comply with the conditions subject to which the certificate has been issued, can revoke the certificate. The certificate can be revoked only after giving a show cause notice to the holder of the certificate.

13

Seed Sampling and Testing

Seed sampling

Seed sampling is to draw a portion of seed from seed lot that represents the entire seed lot. Seed lot is a uniformly blended quantity of seed either in bag or in bulk.

Types of samples

1. Primary sample

Each probe or handful of sample taken either in bag or in bulk is called primary sample.

2. Composite sample

All the primary samples drawn are combined together in suitable container to form a composite sample

3. Submitted sample

When the composite sample is properly reduced to the required size that to be submitted to the seed testing lab, it is called submitted sample. Submitted sample of requisite weight or more is obtained by repeated halving or by abstracting and subsequently combining small random portions.

4. Working sample

It is the reduced sample required weight obtained from the submitted sample on which the quality tests are conducted in seed testing lab.

Sampling intensity

By using appropriate triers, samples can be taken from bags or from bulk. Triers are used for drawing samples from the lots stored in the bins and bags. Based on the number of containers the primary sample size will vary.

a. For seed lots in bags (or container of similar capacity that are uniform in size)

- Up to 5 containers: Sample each container but never < 5 Primary sample
- 6-30: Sample atleast one in every 3 containers but never > than 5 P.S.
- 31-400: Sample atleast one in every 5 containers but never < 10 P.S.
- 401 or more: Sample atleast one in every 7 containers but never < 80.
- When the seed is in small containers such as tins, cartons or packets a 100 kg weight is taken as the basic unit and small containers are combined to form sampling units not exceeding this weight *e.g.*, 20 containers of 5 kg each. For sampling purpose each unit is regarded as one container.

b. For seeds in bulk

- Up to - 500 kg - Atleast 5 Primary sample
- 501 - 3000 Kg - 1 Primary sample for each 300 kg but not less than 5 P.S.
- 3001-20,000 Kg - 1 Primary sample for each 500 kg but not less than 10 P.S.
- 20,001 and above - 1 Primary sample for each 700 kg but not less than 40 P.S.

Sample is obtained from seed lot by taking small portion at random from different places and combining them. From this sample smaller samples are obtained by one or more stages. In each and every stage thorough mixing and dividing is necessary.

Methods of sampling

Hand sampling

This is followed for sampling the non free flowing seeds or chaffy and fuzzy seeds such as tomato, grass seeds etc., In this method it is very difficult to take samples from the deeper layers of bag. To overcome this, bags are emptied completely or partly and then seed samples are taken. While removing the samples from the containers, care should be taken to close the fingers tightly so that no seeds escape.

Sampling with triers

1. Nobbe trier

The name was given after Fredrick Nobbe - father of seed testing. This trier is made in different dimensions to suit various kinds of seeds. It has a pointed tube long enough to reach the centre of the bag with an oval slot near the pointed end. The length is very small. This is suitable for sampling seeds in bag not in bulk.

2. Sleeve type triers or stick triers

It is the most commonly used trier (with compartments and without compartments) for sampling: There are two types *viz.*, It consists of a hollow brass tube inside with a closely fitting outer sleeve or jacket which has a solid pointed end. Both the inner tube as well as the outer tube has been provided with openings or slots on their walls. When the inner tube is turned, the slots in the tube and the sleeve are in line. The inner tube may or may not have partitions. This trier may be used horizontally or vertically. This is diagonally inserted at an angle of 30^0 C in the closed position till it reaches the centre of the bag. Then the slots are opened by giving a half turn in clockwise direction and gently agitated with inward push and jerk, so that the seeds will fill each compartment through the openings from different layers of the bag, then it is again closed and withdrawn and emptied in a plastic bucket. This trier is used for drawing seed samples from the seed lots packed in bags or in containers.

The samples taken may packed in bags, sealed and marked for identification. For moisture testing the samples should be packed separately in moisture proof polythene bag and kept in the container along with the submitted samples.

Mixing and dividing of seeds

The main objective of mixing and dividing of seeds is to obtain the representative homogenous seed sample for analysis by reducing the submitted sample to the desired size of working sample.

Method of mixing and dividing

- Mechanical dividing
- Modified halving method
- Hand halving method
- Random cup method
- Spoon method

1. Mechanical method

- The reduction of sample size is carried out by the mechanical dividers suitable for all seeds except for chaffy and fuzzy seeds.
- To mix the seed sample and make homogenous as far as possible. To reduce the seed sample to the required size without any bias.
- The submitted sample can be thoroughly mixed by passing it through the divider to get 2 parts and passing the whole sample second time and 3rd time if necessary to make the seeds mixed and blended so as to get homogenous seed sample when the same seeds are passed through it into approximately equal parts.
- The sample is reduced to desired size by passing the seeds through the dividers repeatedly with one half remain at each occasion.

Types of mechanical dividers

Boerner divider

It consists of a hopper, a cone and series of baffles directing the seeds into 2 spouts. The baffles are of equal size and equally spaced and every alternate one leading to one spout. They are arranged in circle and are directed inward. A valve at the base of the hopper retains the seeds in the hopper. When the valve is opened, the seeds fall by gravity over the cone where it is equally distributed and approximately equal quantity of seeds will be collected in each spout. A disadvantage of this divider is that it is difficult to check for cleanliness

Soil divider

It is a sample divider built on the same principles as the Boerner divider. Here the channels are arranged in a straight row. It consists of a hopper with attached channels, a frame work to hold the hopper, two receiving pans and a pouring pan. It is suitable for large seeds and chaffy seeds.

Centrifugal or Gamet divider

The principle involved is the centrifugal force which is used for mixing and dividing the seeds. The seeds fall on a shallow rubber spinner which on rotation by an electric motor, throw out the seeds by centrifugal force. The circle or the area where the seeds fall is equally divided into two parts by a stationary baffle so that approximately equal quantities of seed will fall in each spout.

1. Modified halving method

The apparatus consists of a tray into which is fitted a grid of equal sized cubical cups open at the top and every alternate one having no bottom. After preliminary mixing the seed is poured evenly over the grid. When the grid is lifted, approximately half the sample remains on the tray. The submitted sample is successively halved in this method until a working sample size is obtained.

2. Hand halving method

This method is restricted to the chaffy seeds. The seed is poured evenly on to a smooth clean surface and thoroughly mixed into a mound. The mound is then divided into 1/2 and each half is mound again and halved into 4 portions. Each of the 4 portions is halved again giving 8 portions. The halved portions are arranged in rows and alternate portions are combined and retained. The process is repeated until the sample of required weight is obtained.

3. Random cup method

This is the method suitable for seeds requiring working sample upto 10 grams provided that they are not extremely chaffy and do not bounce or roll (*e.g.*) *Brassica* spp. Six to eight small cups are placed at random on a tray. After a preliminary mixing the seed is poured uniformly over the tray. The seeds that fall into the cup is taken as the working sample.

4. Spoon method

This is suitable for samples of single small seeded species. A tray, spatula and a spoon with a straight edge are required. After preliminary mixing, the seed is poured evenly over the tray. The tray should not be shaked thereafter. With the spoon in one hand, the spatula in the other and using both small portions of seed from not less than 5 random places on the tray should be removed. Sufficient portions of seed are taken to estimate a working sample approximately but not less than the required size.

Types of samples submitted in seed testing laboratory

Certified sample

Samples drawn by seed certification officers from crops offered for seed certification. The analytical fee per sample is collected by the Assistant Director of Seed Certification. Tests conducted are moisture test, physical purity, germination test and ODV test.

Official sample

Samples drawn by Seed Inspectors in different places like seed selling points and production centres under seed law enforcement programme. No analytical fee is collected as it comes under law enforcement. Tests conducted are physical purity and germination test.

Service sample

The samples sent by seed producers, dealers and farmers. The analytical fee is per sample was collected. This amount can be sent by M.O. (or) personal DD. Test on germination alone is conducted.

Seed Testing

Seed testing is the science of evaluating the planting value of seed. It is determining the standards of a seed lot *viz*., physical purity, moisture, germination and ODV and thereby enabling the farming community to get quality seeds. The Seed Testing Laboratory is the hub of seed quality control. Seed testing services are required from time to time to gain information regarding planting value of seed lots. Seed testing is possible for all those who produce, sell and use seeds.

Objectives of seed testing

Seed testing is required to achieve the following objectives for minimizing the risks of planting low quality seeds.

- To identify the quality problem and their probable cause
- To determine their quality, that is, their suitability for planting
- To determine the need for drying and processing and specific procedures that should be used
- To determine if seed meets established quality standards or labeling specifications.
- To establish quality and provide a basis for price and consumer discrimination among lots in the market.
- The primary aim of the seed testing is to obtain accurate and reproducible results regarding the quality status of the seed samples submitted to the Seed Testing Laboratories.

Role of Seed Testing Laboratories

Seed testing laboratories are essential organization in seed certification and seed quality control programmes. The main objective is to serve the producer, the consumer and the seed industry by providing information on seed quality. Test results may cause rejection of poor seed multiplication or low grade seed in a count of law.

Analysis of seed in the laboratory

Seed testing is possible for all those who produce, sell and use seeds. Seed testing is highly specialized and technical job. With a view to maintain uniformity in quality control the seed analysis laboratory includes for distinct sections.

- Section for purity testing.
- Section for moisture testing.
- Section for viability, germination and section for vigour testing.

14

Purity Analysis and Moisture Estimation

Purity analysis

Purity of a seed lot indicates in percentage how large a fraction is made up of pure seeds of the species in question, and how much is made up of inert matter and other seeds. Impurities may be any non seed material (leaf, flower, fruit fractions, soil etc.), small fractions of seeds of the actual species, as well as seeds of other species. Pure seed may include both dead and empty seed, plus damaged seed, purity does not tell anything about viability. The purity analysis of a seed sample in the seed testing laboratory refers to the determination of the different components of the purity *viz.,* pure seeds, other crop seeds, weed seeds and inert matter. Each component is reported as a percentage of total weight. The objective of the purity analysis is to determine whether the submitted sample conforms to the prescribed physical quality standards with regard to physical components. Large seeds are generally easy to clean and purity analyses are therefore often omitted. However, for smaller seeds contamination with other seeds may occur *e.g.*, during processing.

A purity analysis at a certain stage of processing may serve as a guideline for the necessity of further cleaning. However, in practice never possible to achieve a completely clean or pure seed lot by mechanical processing, because the physical characters of other seed or inert matter may be so similar to those of the seeds in question that separation is impossible. Yet information on the actual composition of the seed lot is important for the seed handler, both in relation to the factors of seed price and seed demand and because any type of impurity may hamper practical sowing. Certain types of impurity may also harbour infective fungi, which in turn could hamper seed quality.

The working sample

The purity analysis is done on the working sample of prescribed weight drawn from submitted sample. The analysis may be made on one working sample of the prescribed weight or on two sub-samples of atleast half of this weight, each independently drawn.

Weighing the working sample

The number of decimal places to which the working sample and the components of the working sample should be weighed is given below.

Weight of the working sample (g)	The number of decimal places required	Example
<1	4	0.7534
1-9.999	3	7.534
10-99.99	2	75.34
100-999.9	1	753.4
1000 or more	0	7534

Purity separation

The working sample after weighing is separated into its components *viz.,* pure seed, other seed crop, weed seed and inert matter.

- **Pure seed -**The seeds of kind/species stated by the sender. It includes all botanical varieties of that kind/species. Immature, undersized, shrivelled, diseased or germinated seeds are also pure seeds. It also includes broken seeds, if the size is >1/2 of the original size except in leguminaceae and cruciferae where the seed coat entirely removed are regarded as inert matter.
- **Other crop seed-**It refers to the seeds of crops other than the kind being examined.
- **Weed Seed-** It includes seeds of those species normally recognized as weeds or specified under Seed Act as a noxious weed.
- **Inert matter-** It includes seed like structures, stem pieces, leaves, sand particles, stone particles, empty glumes, lemmas, paleas, chaff, awns, stalks longer than florets and spikelets.

Method of purity separation

Place the sample on the purity work board after sieving / blowing operations and separate into other crop seeds and inert matter. After separation, identify each kind of weed seeds, other crop seeds as to genus and species. The names and number of each are recorded. The type of inert matter present should also be noted.

Calculation

All the four components must be weighed to the required number of decimal places. The percentages of the components are determined as follows.

$$\% \text{ of components} = \frac{\text{Weight of individual component}}{\text{Total weight of all components}} \times 100$$

If there is a gain or loss between the weight of the original samples and the sum of all the components is in excess of one percent, another analysis should be made.

Duplicate tests

If the analysis result is near the border line in relation to the seed standards, one more test is done and the average is reported. However, if a duplicate analysis is made of two half sample or whole samples, the difference between the two must not exceed the permissible tolerance. If the difference is in excess of the tolerance, analyze further (but not more than 4 pairs in all) until a pair is obtained which has its member within tolerance.

Moisture estimation

The moisture content of a seed sample is the loss in weight when it is dried. It is expressed as a percentage of the weight of the original sample. It is one of the most important factors in the maintenance of seed quality. Moisture content of seeds tends to vary with atmospheric humidity, it is important that exposure to varying humidity is minimized before testing. Therefore, seeds should be packed in water proof material as quickly as possible after sampling. In order to avoid the possibility of water condensing on the seed when removed from cold store, seed should be allowed to reach ambient temperature before the container is opened. Under laboratory testing, seed moisture is measured by the oven drying method, which is the direct method, prescribed by ISTA. This method can also be used for calibrating moisture meters for indirect measurement of moisture content. The indirect methods provide very quick results, which can be used as a guide during seed handling, *e.g.*, to determine the necessity for further drying.

Method of moisture determination

- Direct method
- Indirect method

Direct method

Under this category, the seed moisture content is measured directly by loss or gain in seed weight. These are:

- Desiccation method
- Phosphorus pentaoxide method
- Oven-drying method
- Vacuum drying method
- Distillation method
- Karl Fisher's method
- Direct weighing balance
- Microwave oven method

Oven drying method

The oven drying method is broadly grouped into two categories:

- Low Constant Temperature Oven Method
- High Constant Temperature Oven Method

Low constant temperature oven method

This method has been recommended for seed of the species rich in oil content or volatile substances. In this method, the pre- weighed moisture bottles along with seed material are placed in an oven maintaining a temperature of 103°C. Seeds are dried at this temperature for 17+-1hr.The relative humidity of the ambient air in the laboratory must be less than 70 per cent when the moisture determination is carried out.

High constant temperature oven method

The procedure is the same as above except that the oven is maintained at a temperature of 130°-133°C. The sample is dried to a period of one hour. In this method there is no special requirement pertaining to the relative humidity of the ambient air in the laboratory during moisture determination.

Essential equipments and supplies

- Constant temperature precision hot-air electric oven
- Weighing bottles/Moisture containers

- Desiccator with silica gel
- Analytical balance capable of weighing up to 1mg
- Seed grinder/An adjustable grinding mill
- Tong
- Heat resistant gloves
- A brush/A steel brush

Sample size

The ISTA rules recommend that two replicates, each with 4 gm of seed be used for determination of seed moisture content. This seed sample weight may be modified to 0.2 to 0.5 gm per replicate, with precise weighing, for use in seed gene banks, to avoid unnecessary depletion of precious biological resources.

Procedure

- Seed moisture determination is carried out in duplicate on two independently drawn working samples.
- Weigh each bottle with an accuracy of 1 mg or 0.1 mg.
- First weigh the empty bottle/container with its cover.
- Grind the seed material, evenly using any grinder/grinding mill that does not cause heating and/or loss of moisture content.
- Mix thoroughly the submitted sample, using spoon, and transfer small portions (4 to 5 gm) of seed samples directly into weighing bottles/containers, by even distribution on bottom of the containers.
- After weighing, remove the cover or lid of the weighing bottles/containers.
- Place the weighing bottles/containers in an oven, already heated to or maintaining the desired temperature, for the recommended period.
- At the end of seed drying period, weighing bottles/containers be closed with its lid /cover.
- Transfer the weighing bottles/containers to the desiccators having silica gel (self indicating -blue), to cool down for 40 to 45 min.
- Weigh again the cooled weighing bottles/containers.
- Calculate the seed moisture content.

Calculation of results

The moisture content as a percentage by weight (fresh weight basis) is calculated to one decimal place, by using of the formulae:

$$\text{Percentage seed moisture content (mc)} = \frac{M2\text{-}M3}{M2\text{-}M1} \times 100$$

Where

M1 = Weight of the weighing bottle/container with cover in gm

M2 = Weight of the weighing bottle/container with cover and seeds before drying

M3 = Weight of the weighing bottle/container with cover and seeds after drying

{Note: The seed moisture determination must be done in two replicates, with precise weighing (*i.e.*, up to three decimal places) using lightweight weighing bottles/containers.}

If the seed is pre-dried or dried in two steps: The seed moisture content is calculated from the results obtained in the first (pre-dried) and second stages of seed drying, using the following formula, and expressed as percentages, as under:

$$\text{Percentage seed moisture content (mc)} = \frac{(S1 + S2) - (SI * S2)}{100}$$

Where

S1= is the moisture loss in the first stage, and

S2= is the moisture loss in the second stage

Indirect method

These are no so accurate; estimation is approximate, but convenient and quick in use. These are frequently used at seed processing plants. These measure other physical parameters like electrical conductivity or electrical resistance of the moisture present in the seed. Values are measured with the help of seed moisture meters, and these values are transformed into seed moisture content with the help of calibration charts, for each species, against standard air-oven method or basic reference method.

Universal (OSAW) digital moisture meters

The principle involved in these moisture meters is that wet grains are good conductors while dry grains are less conductors of electricity. So, the moisture content is directly proportional to the electrical conductivity of the seed. It consists of a compression unit to compress the sample to pre -determined thickness. The thickness setting is very easily read on a vertical and circular scale. The seed material on test is taken in a test cup and is compressed. Then press the push type switch till the reading comes in the display. Here no temperature reading and correlated dial are required. The computer version of digital moisture meter automatically compensate for temperature corrections.

Above all Karl-Fisher's method has been considered as the most accurate and the basic reference method for standardizing other methods of seed moisture determination. The constant temperature oven drying method is the only practical method, approved by International Seed Testing Association (ISTA) and other organization to be used for routine seed moisture determination in a seed-testing laboratory.

Use of tolerances

Result is the arithmetic mean of the duplicate determination of seed moisture content, for a given seed sample. The maximal difference of 0.2% is recommended between two replicates, for crop seed species under1STA rules. If the difference between two replicates exceeds 0.2%, the seed moisture determination in duplicate is repeated. As it is very difficult, rather impossible to, meet the replicate difference of seed moisture up to 0.2% in tree or shrub species, maximal limit of 0.3 to 2.5% is recommended between two replicates for seed moisture in tree or shrub species under ISTA rules.

Reporting of results

Seed moisture content is reported to the nearest 0.1% on ISTA analysis certificate. If the seed moisture content is determined using any moisture meter, the brand name and type of the equipment be mention on the analysis certificate, under column of "other determinations" reporting of range for which the moisture meter is calibrated is the another requirement, on seed analysis certificate.

15

Seed Germination Test

Seed germination

It is defined as the emergence and development from the seed embryo, of those essential structures, for the kind of seed in question, indicates its ability to produce a normal plant under favourable conditions. The seedlings devoid of an essential structure; showing weak or unbalanced development; decay or damage affecting the normal development of seedling are not considered in calculating the germination percentage.

Types of germination

1. Epigeal germination

During germination the cotyledons are raised above the ground. During root establishment the hypocotyl begins to elongate in an arch which breaks thro' the soil, pulling the cotyledon and enclosed (epicotyl) thro' the ground and projecting them into the air. (*e.g.*) bean, castor, cucurbits and other dicots and onion.

2. Hypogeal germination

During germination, cotyledons remain beneath the soil while the plumule pushes upward and emerges above the ground. Here the epicotyl (plumule) elongates (*e.g.*) Peas, grams, mango, grasses and many other spp.

Principles of seed testing

The ultimate aim of testing the germination in seed testing laboratory is to obtain information about the planting value of the seed sample and by inference the quality of the seed lot. In addition, the laboratory germination results are also required for comparing the performance potential or superiority of the different seed lots. In general, the farmers, seeds men and public agencies use the germination results for the following purposes:

- Sowing purposes, with a view to decide the seed rate to achieve desired field establishment
- Labelling purposes
- Seed certification purposes
- Seed Act and Law Enforcement purposes.

1. Method of testing

The first and foremost step is to draw a true representative sample from the seed lot. To obtain a random sample for testing it is always best to take samples from different parts of the bag or container. If the seed to be tested is from a seed lot that contains more than one bag, samples must be taken from several bags. The sample thus drawn is further divided and the required numbers of seeds are the taken to perform the actual test.

2. Essential equipments

The following pieces of equipments are essential to carry forward the germination tests in the seed testing laboratories.

a) Seed Germinator

It is the essential requirement for germination testing for maintaining the specific conditions of temperature, relative humidity and light. The seed germinators are generally of two types namely: Cabinet germinator and walk in germinator. The cabinet seed germinators are essential under the situations, where various kinds of seeds that require different sets of conditions, are being handled in the laboratory. The number of the pieces of the germinators required by the laboratory will depend on the number of seed samples and the species being analyzed by the laboratory. The seed testing laboratories that handle large number of seed samples and require maintaining only fewer (2-3) sets of temperature conditions, the walk-in-germinators are preferred. Such germinators are more useful for conducting the germination tests in sand media, which require large germination space.

b) Counting devices

The counting devices include the counting boards, automatic seed counter and vacuum seed counter. These devices are required to aid germination testing by minimizing the time spent on planting the seeds as well as to provide proper spacing of the seed on germination substrata. Counting boards are suitable for medium and bold sized seeds, while vacuum counter can be, used for small sized seeds. In the absence of counting devices, the work may be accomplished manually.

c) Other equipments

The other equipments required for germination testing include the refrigerators, scarifier, hot water bath, incubator, forceps, spatula, germination boxes, plastic plates, roll- towel stands and plastic or surgical trays, etc. A large oven with temperature range 100-200°C is also required for sterilizing the sand.

d) Miscellaneous - Glassware and Chemicals

Germination paper (Creppe Kraft paper or towel paper, sunlit filter paper and blotters) and sand are the basic supplies required for germination tests. In addition, the laboratory may also require some glassware, such as petridishes, beakers, funnel, measuring cylinders, muslin cloth, rubber bands and tubes etc. and certain chemicals like Potassium nitrate, Thiourea, Gibrellic acid, and Tetrazolium chloride for specific purposes. Voltage stabilizers are required for the supply of the constant electric current. The voltage stabilizers are essential for costly germinators air-conditioners and refrigerators. Under the situations of erratic power supplies and breakdowns, electricity generators are also required.

3. Care of equipments

The seed analyst must ensure that

- All the equipments are in proper working condition
- The germinators are maintaining correct temperature and relative humidity inside the germinator is maintained 90-98%
- The phytosanitary conditions of the germinators and germination trolleys are adequate
- The germinators are disinfected periodically by flushing with hot water or solution of

Potassium permanganate or chlorine water

- The temperature and the relative humidity of the walk-in-germinators are recorded daily and displayed on a chart
- The floor, ceiling and walls of the Walk-in-germinator are devoid of cracks, crevices.

Evenly plastered and duly painted to avoid contamination by fungus, bacteria or insects.

4. Handling of substrata

The accuracy and reproducibility of the germinator result are very much dependent on the quality of the substrata (paper and sand) used for germination testing. The germination substrata must meet the following basic requirements:

- It should be non-toxic to the germinating seedlings
- It should be free from moulds and other microorganisms
- It should provide adequate, aeration and moisture to the germinating seeds
- It should be easy to handle and use
- It should make good contrast for judging the seedlings
- It should be less expensive

a) Paper substrata

The paper substrata are used in the form of top of paper (TP) or between paper (BP) tests. In most of the laboratories, paper-toweling method (Roll towel test) is most commonly used for medium sized and bold seeds. The paper substrata are not reusable.

b) Sand substrata

The sand substrata have advantage of being relatively less expensive and reusable. The results in sand media are more accurate and reproducible in comparison with 'roll towel' tests. The sand should be reasonably uniform and free from very small and large particles. It should not contain toxic substances and its pH should be within the range of 6.0- 7.5. The sand should be washed, sterilized and graded with a sieve set having holes of 0.8 mm diameter (upper sieve) and 0.05mm diameter (bottom sieve). The sand retained on the bottom sieve should only be used.

5. Testing of Substrata

a) Phytotoxicity

The substrata should be tested for its phytotoxicity, capillary rise, moisture holding capacity and bursting strength, etc., before accepting the supplies in the laboratory. Periodic checks of the quality of the substrata should also be made in the laboratory. By germinating the seeds of *Brassica*, Onion, Chillies or Berseem and studying the phytotoxic symptoms on the germinating seedlings can check the phytotoxicity of the paper or sand substrata. The phytotoxic symptoms include shortened roots; discoloured root tips; root raised from the paper; inhibition of root hairs development and root hairs bunched. The symptoms are more pronounced at an early stage of root growth. The phytotoxic symptoms are also evident in the plumular areas in the form of thickened or flattened plumules or coleoptiles.

b) Capillary Rise

Cut four strips of germination paper10 mm wide; two in machine direction and the other two in cross machine direction. Take distilled water in small glass beakers. Immerse one end of each strip in the water to depth of 20 mm. Wait for 2 minutes and then measure the height to which water has risen in the strip to the nearest mm. Compute the average for the two strips cut in machine direction or cross machine direction separately. The lower value of the two averages should be considered as capillary rise.

c) Bursting strength

The bursting strength of the paper is measured with equipment however, it can be checked as follows:

- Hold the two ends of the germination paper and exert the pressure by stretching the paper with mid force.
- Soak the paper in water for 1-2 hours.
- The paper of desired bursting strength would not tear off easily

6. Test conditions

a) Moisture and aeration

The moisture requirements of the seed will vary according to its kind. Large seeded species require more water than the small seeded species. It is essential that the substratum must be kept moist throughout the germination period. Care need to be taken that the sub- stratum should not be, too moist. The excessive moisture will restrict the aeration and may cause the rotting of the seedlings or development of watery seedlings. The water used for moistening the substratum must be free from organic and inorganic impurities. Normally the tap water is used. The pH of the water is not satisfactory, distilled water or deionized water may be used. Under such situation care need to be exercised to aerate the tests frequently to provide oxygen supply to the germinating seedlings because oxygen level in distilled water is very low. The initial quantity of water to be added to the substratum will also depend on its nature and dimensions. Subsequent watering, if, any may be left to the discretion of the analyst but it should be avoided as far as possible because it may cause the variation in germination results. In order to reduce the need for additional watering during the germination period, the relative humidity of the air surrounding the seeds should be kept at 90-95 % to prevent loss of water by evaporation. Special measures for aeration are not usually necessary in case of top of paper (TP) tests. However, in case of 'Roll towel' tests (BP) care should be taken that the rolls should be loose enough to allow the presence of sufficient air around the seeds. In case of sand media, the sand should not be compressed while covering the seeds.

b) Temperature

At very low or high temperatures, the germination is prevented to a larger extent. The temperature should be uniform throughout the germinator and the germination period. The variation in temperature inside the germinator should not be more than 1DC. The prescribed temperature for germination of agricultural, vegetable or horticultural seeds, provided in the Rules for Seed Testing can be broadly is classified into two groups, *viz.*, constant temperatures and alternate temperatures.

Constant temperature: The tests must be held at the specific temperature during the entire germination period.

Alternate temperature: The lower temperature should be maintained for 16 hours and the higher for 8 hours; a gradual change over lasting 3 hours is usually satisfactory for non-dormant seeds. However, a sharp changeover lasting 1 hour or less, or transfer of test to another germinator at lower temperature may be necessary for seeds, which are likely to be dormant.

c) Light

It is always better to illuminate the tests for the proper growth of the seedlings. Under the situations where light is essential for germinations, tests should be exposed to the natural or "artificial source of light. Seeds that require light for germination must be illuminated with cool fluorescent light for at least 8 hours in every 24 hours cycle. Under the situation where testing of the seed is required to be undertaken alternating "temperatures together with light, the tests should be illuminated during high temperature period.

7. Laboratory procedures

Four replication of 100 seeds, A minimum of 3 replication of 100 seeds may be used under unavoidable situations or Eight or six replications of 50 seeds or Sixteen/twelve replication of 25 seeds according to the kind of and size of containers.

8. Methods to improve germination

- **Hard seeds:** Seeds with hard seed coats may germinate more readily after soaking for up to 24-48 hours in water, Careful-piercing, chipping, filing or sand papering of the seed coat may be sufficient to break the dormancy condition. Care must be taken to scarify the seed coat at a suitable part in order to avoid damaging the embryo. The best site for mechanical scarification is that part of the seed coat immediately above the tips of the cotyledons. Treating with in concentrated Sulphuric acid (H_2SO_4) is effective with some species.
- **Inhibitory Substances:** Naturally occurring substances in the pericarp or seed coat, which act as inhibitors of germination may be removed by washing the seeds in running water at a temperature of 25°C before the germination test is made. After washing, the seeds should be dried back at a maximum temperature of 25°C.
- **Disinfection of the seed:** For samples of *Beta vulgaris*, a fungicide treatment may be applied before planting the seed for germination.
- **Prechilling:** In some seeds having physiological dormancy pre chilling is required for inducing germination. Replicates for germination are placed in contact with the moist substratum and kept at a low temperature for an initial period. Vegetable seeds are kept at a temperature between 5°C and 10°C for an initial period up to, 7 days. Tree seeds are kept it a temperature between3°C and 5°C, for a period, varying with the species, from 7 days to 12 months.
- **Pre-drying:** The replicates for germination should be heated at a temperature not exceeding 40°C with free air circulation for a period of up to 7 days before they are placed under the prescribed germination conditions. In some cases it may be necessary to extend the pre-drying period. Both the duration and the temperature should be reported on the Analysis Certificate.
- **Chemical Treatments**: Chemicals are used for breaking the dormancy. *e.g.*, Potassium nitrate, Gibberellic acid.

9. Duration of testing

The duration of the test is determined by the time prescribed for the, final count. If at the end of the prescribed test period some seeds have just started to germinate, the test may be extended for an additional period up to 7 days. A test may be terminated prior to the prescribed time when the analyst is satisfied that the maximum germination of the sample has been obtained. The time for the, first count is approximate and a deviation of 1-3 days is permitted. The first count may be delayed to permit the development of root hairs in order to be certain that root development is normal, or may be omitted. Intermediate counts may be at the discretion of the analyst to remove seedlings, which have reached a sufficient state of development for evaluation, to prevent them becoming entangled. But the number of intermediate counts should be kept to a minimum to reduce the risk of damaging any seedlings that are not sufficiently developed. Seedlings may have to be removed and counted at more frequent intervals during the prescribed period of the test when a sample contains is infected with 'fungi or bacteria. Seeds that are obviously dead and decayed, and may, therefore, be a source of contamination for healthy seedlings, should be removed at each count and the number recorded.

10. Evaluation of germination test

The germination tests need to be evaluated on the expiry of the germination period, which varies according to the kind of seed. However, the seed analyst may terminate the germination test on or before the final count day or extend the test beyond the period depending on the situation. First and second counts are usually taken in case of Top of Paper (TP) and Between Paper (BP) media; however a single final count is made in case of and tests. At the first and subsequent counts only normal and dead seeds (which are source of infection) removed and recorded. In evaluating the germination test the seedling and seeds are categorized into Normal seedlings, abnormal seedlings, dead seeds, fresh un germinated and hard seeds. It may also be necessary to remove the seed coat and separate the cotyledons In order to examine the plumule in species where essential structures are still enclosed at the end of the test.

Categories of seedlings

- Normal seedlings
- Abnormal seedlings
- Hard seeds
- Fresh ungerminated seeds

i. Normal seedlings

Characters of Normal Seedlings

- A well developed root system with primary root except in certain species of graminae which normally producing seminal root or secondary root.

- A well developed shoot axis consists of elongated hypocotyl in seedlings of epigeal germination.
- A well developed epicotyl in seedlings of hypogeal germination
- One cotyledon in monocotyledons and two in dicotyledons
- A well developed coleoptile in graminae containing a green leaf.
- A well developed plumule in dicotyledons.
- Seedlings with following slight defects are also taken as normal seedlings. Primary root with limited damage but well developed secondary roots in leguminosae (*Phaseolus*, *Pisum*), Graminae (maize), Cucurbitaceae (Cucumis) and Malvaceae (cotton).
- Seedlings with limited damage or decay to essential structures but no damage to conducting tissue.
- Seedlings which are decayed by pathogen but it is clearly evident that the parent seed is not the source of infection.

ii. Abnormal seedlings

Seedlings which do not show the capacity for continued development into normal plant when grown in favourable conditions of soil, water, temperature and light.

Type of abnormal seedlings

A. Damaged seedlings

Seedlings with any one of the essential structures missing or badly damaged so that the balanced growth is not expected. Seedlings with no cotyledons, with splits, cracks and lesions or essential structures and without primary root.

B. Deformed Seedlings

Weak or unbalanced development of essential structures such as spirally twisted or stunted plumule or hypocotyl or epicotyl, swollen shoot, stunted roots etc.,

C. Decayed seedlings

Seedlings with any one of the essential structures showing diseased or decayed symptoms as a result of primary infection from the seed which prevents the development of the seedlings.

iii. Hard seeds

Seed which do not absorb moisture till the end of the test period and remain hard (e.g) seeds of leguminosae and malvaceae.

iv. Fresh ungerminated seeds

Seeds which are neither hard nor have germinated but remain firm and apparently viable at the end of the test period.

v. Dead seeds

Seeds at the end of the test period are neither hard nor fresh or have produced any part of a seedling. Often dead seeds collapse and milky paste comes out when pressed at the end of the test.

11. Retesting

If the results of a test are considered unsatisfactory it shall not be reported and a second test shall be made by the same method or by alternative method under the following circumstances.

- Replicates performance is out of tolerance
- Results being inaccurate due to wrong evaluating of seedlings or counting or errors in test conditions.
- Dormancy persistence or phytotoxicity or spread of fungi or bacteria. The average of the two tests shall be reported.

12. Use of tolerances

The result of a germination test can be relied upon only if the difference between the highest and the lowest replicates is within accepted tolerances. To decide if two test results of the same sample are compatible again the tolerance table is used.

13. Reporting results

The results of the germination test are calculated as the average of 4 x 100 seed replicates. It is expressed as percentage by number of normal seedlings. The percentage is calculated to the nearest whole number. The percentage of abnormal seedlings, hard, fresh and dead seeds is calculated in the same way. These should be entered on the analysis of certificate under appropriate space. If the result is nil for any of these categories it shall be reported as 'O'.

Crop	Substratum	Temp (°C)	First count days	Final count days	Pre-treatment
Paddy	BP, TP, S	20-30	5	14	Preheat (50°C) soak in H_2O or HNO_3 24hrs
Maize	BP, S	20-30	4	7	-
Bajra	TP, BP	20-30	3	7	0.2% KNO_3 (2-3 hrs) pre chill
Sorghum	TP, BP	20-30	4	10	-
Redgram	BP, S	20-30	4	6	-
Black gram	BP, S	30	4	7	-
Green gram	BP, S	20-30	5	8	-
Bengalgram	BP, S	20-30	5	8	-
Cowpea	BP, S	20-30	5	8	-
Peas	BP, S	20	5	8	-
Castor	BP, S	20	7	14	
Groundnut	BP, S	20-30	5	10	-
Sunflower	BP, S	20-30	4	10	-
Sesame	TP	20-30	3	6	-
Cotton	BP, S	20-30	4	12	Remove shells
Brinjal	TP, BP	20-30	7	14	Ethrel (25ppm) 48hrs
Tomato	TP, BP	20-30	5	14	-
Chillies	TP, BP	20-30	7	14	Hot water 85°C 1min.
Bhendi	BP, S	20-30	4	21	-
Onion	TP, BP	15-20	6	21	KNO_3
Carrot	TP, BP	20-30	7	14	KNO_3
Radish	TP, BP	20-30	4	10	Pre chill
Cauliflower	TP	20-30	5	10	Pre chill, KNO_3
Ashgourd	S	30-35	5	14	-
Bitter gourd	BP, S	20-30	4	14	-
Bottle gourd	BP, S	20-30	4	14	-

16

Seed Viability and Vigour Tests

Seed viability tests

Only two methods, the Topographical tetrazolium (TZ) test and the embryo excision test were previously accepted by the International Seed Testing Association as official methods for some species of seeds. ISTA has recently accepted the X-ray method as a valid alternative to the cutting test for the detection of empty and insect-damaged seeds. The following tests can be applied depending on the circumstances.

1. Quick viability test

The relative long periods of time required for completion of germination tests delays the seed marketing. This necessitated the development of rapid methods for estimating the germination capacity of seeds. This test was developed by Lakon (1942) in Germany.

Principle

It is a biochemical test, in which living cells are made visible by reduction of an indicator dye. The indicator used is 2,3,5triphenyl tetrazolium chloride. Within the seed tissues, it interferes with the reduction processes of living cells and accepts hydrogen from the hydrogenases. By hydrogenation of the 2,3,5 - tri phenyl tetrazolium chloride; a red stable and non diffusible substance, water-insoluble compound triphenyl formazan is produced in living cells.

Procedure

- This test is not valid for previously germinated seeds.
- Seeds are soaked in water overnight. They may be pre-moistened, in which case the seeds are allowed to imbibe water between a moistened germination paper blotter.

- Seeds are then dissected, either longitudinally or transversely, with a scalpel so that the embryo is exposed to the tetrazolium chloride solution. One half of this seed is used for the test and the other half is discarded.
- A solution of 2, 3, 5 triphyenyl tetrazolium chloride (a salt) is added to water to form a colourless solution.
- A representative sample of 50 (or) 100 seeds is usually sufficient. However, 200 seeds, in replicates of 100 seeds are recommended.
- The seeds are placed in a 1% solution (for legumes and grasses that are not bisected), or a dilute 0.1% solution for bisected grasses and cereals.
- The staining time varies for different kinds of seeds, different methods of preparation, and different temperatures (< 1 hr to 8 hrs).
- Seed coats of legumes must usually be removed or peeled back before examination. Care must be taken to prevent breaking of radicles and other damage to the seeds.
- Seed samples with water can also be kept for 3 days at 10°C for interpretation.

Evaluation

Depending on size, all seeds are examined under a microscope at 10-30 power. Larger seeds, such as peas, may be examined without a microscope. Analysts look for three things:

- Sound tissues produce a normal red colour and resist the penetration of tetrazolium. The rate of hydrogen released in sound tissue is slow in comparison to that in partially weakened tissues.
- Weak living tissues produce an abnormal colour. These tissues have lost some of their initial resistance to the penetration of tetrazolium. Respiration is accelerated and formazan is produced rapidly. During the early stages of deterioration, these tissues become darker red (bruised) quicker than sound, healthy tissues.
- Dead tissues do not stain, remaining usually white (aged tissue) because the lack of respiration prevents the production of formazan.

An accurate interpretation of the TZ test depends on

- Knowledge of seed and seedling structure and seed germination.
- Understanding of individual embryo structures and their interdependence in the development of respective seedling structures.

- Understanding of the mechanism of the test and its limitations.
- Combining interpretation of staining patterns with other visible aspects of seed quality.
- Experience with making comparative germination test.

Advantages of TZ test

- Quick estimation of viability
- When the seed is dormant, the TZ test is extremely useful
- Seeds are not damaged (in dicot) in analysis therefore they could be germinated.

Disadvantages of TZ Test

- It is difficult to distinguish between normal and abnormal seedlings.
- It does not differentiate between dormant and non- dormant seeds.
- Since the TZ test does not involve micro organisms harmful to germinating seedlings are not detected.

Reporting results

- TZ test results are recorded as a percentage and are usually reported in the remarks section of the Certificate of Seed Analysis issued by the seed laboratory.
- Hard seeds are to be reported (if applicable) as a percentage and are included in the total percent viable seed.

Accuracy

Properly conducted TZ and germination test results are generally in close agreement and within the range of normal sampling variation. Differences of 3% to 5% may be due to an unavoidable sampling variation error. Differences between TZ and germination test results are usually smaller in high quality seed than in low quality seed, with large seeded crops than with small-seeded crops, and with uniform seed lots than with non-uniform lots.

2. Cutting test

The simplest viability testing method is direct eye inspection of seeds which have been cut open with a knife or scalpel. If the endosperm is of normal colour with a well developed embryo, the seed has a good chance of germinating. This test is not very reliable. Seeds with milky, unfirm, mouldy, decayed, shrivelled or rancid-smelling embryos and abortive seeds that have no embryo can be judged as non-viable without much difficulty. But it is not possible to distinguish moribund, recently dead or recently injured seeds which still appear the same as sound ones. The cutting test is used for ungerminated seeds; it is also a useful tool in estimating the size and maturity of the seed crop before collection and the efficiency of methods used in processing.

3. Excised embryo test

In this method, the seeds are soaked for 1-4 days and the embryos are then excised from the seeds and placed on moist filter paper or blotter discs in petridishes. The tests are placed in the light at a constant temperature of 20°C. The condition of the embryos is examined daily. Depending upon the species and lot differences, the tests can be terminated after only a few days, up to a maximum of 14 days, or as soon as distinct differentiation into viable and non-viable embryos can be made. The excised embryo test is similar to germination tests in that it measures the quality of the seed by their actual germination. In addition it allows some measures of the embryo dormancy to be made, by counting those seeds which, although not growing normally, have grown slightly, remained firm and have kept their colour for the test period. The test is not valid for previously germinated seeds and must not be applied to samples which contain any dry germinated seeds. The success of the test requires considerable skill and experience in the operator and the ISTA rules restrict it to only a few species.

4. Radiographic methods

Radiography was first used to determine seed quality over 70 years ago. The X-ray method permits the detection of empty seeds, mechanical damage and abnormally developed internal seed structures, measurement of the thickness of the seed coat and assessment of the seed viability when combined with a contrast agent. The X-ray contrast method is based on the principle of semi permeability. When seeds are treated with a contrast agent, for example aqueous $BaCl_2$ or vaporous $CHCl_3$, their living tissues are able to prevent its entry due to their semi-permeability, but the dead tissues become impregnated. The impregnated tissues absorb X-radiation more intensively than the unimpregnated ones and thus appear lighter on the film than the unimpregnated ones. The contrast permits living and dead tissue to be located in the seed and an estimation of its viability. The development of soft X-ray equipment has greatly simplified the operation. Complicated photographic equipment is not necessary and pictures can be made with polaroid film which provides clear and detailed radiographs within 30 seconds. A technique of stereo radiography as a supplement to the X-ray contrast method for use in seed quality testing has been developed by Kamra, Meyer and Wegelius (1973). The chief advantage of stereo radiography is that it is possible for the observer to have a three-dimensional view of the object from a pair of radiographs. In this way, the exact topographical location of the contrast agent in the seed can be reliably determined. This increases the information which can be obtained from radiographs and adds to the analytical accuracy.

The X-ray method is a useful one and is likely to play an increasing role in seed testing. Earlier models of X-ray machines were expensive, but recent models, particularly from Japan are much cheaper and now cost less than a cabinet germinator. Improvements in photographic films and paper have speeded up the process and simplified the interpretation, so that technicians can be trained easily to produce consistent results. ISTA has accepted the method as a valid alternative to the cutting test for the detection of empty and insect damaged seeds. It also shows considerable promise for distinguishing between viable and non-viable seeds among "full seeds".

5. Hydrogen peroxide

Hydrogen peroxide (H_2O_2) has a stimulating effect on seed germination and has been used in a rapid test for germination of several conifers in the western USA. Seeds are soaked overnight in 1% H_2O_2. The seed coat is then cut open to expose the radicle tip and the seeds put back into 1% H_2O_2 in the dark at alternating temperatures (20° and 30°C). Counting and refreshment of H_2O_2 is done after 3 or 4 days and final assessment after 7 or 8 days. Radicle growth of 5 mm or more is scored "evident", 0-5 mm "slight" and no growth means a nonviable or empty seed. The test is quicker but less reliable than a normal germination test (usually producing a more rapid and higher final germination), slower but simpler to perform than excised embryo and easier to interpret than TZ.

Seed vigour tests

A vigour test reveals a seed lot's ability to withstand a variety of different stress factors. Vigour tests are designed to mimic poor seeding conditions to find out how the seed lot will perform under stress. It's the exact opposite of a germination test, where seed is grown under optimum conditions. In a vigour test, the seed is introduced to a stressful environment unfavourable to seedling development. This environment can be cool, cold or warm, or a combination of either high humidity and high temperatures, or heavy moisture at low temperatures. If the seed lacks vigour, one or more of these created stressors will suppress seedling growth but, if the seed is vigorous, it will withstand one or all of these stressors and grow as if it were on stimulants. Taken together, vigour and germination test results provide a complete performance profile for a wide range of field conditions and can help guide important seeding decisions, such as when to seed: should you wait for warmer soil temperatures or can this seed survive an earlier planting into cooler soils. Vigour testing is an important component of seed testing because it's more sensitive test than germination, and because loss of vigour may be noted much earlier than loss of germination.

Definitions

ISTA:"Seed vigour is the sum of those properties which determine the potential level of activity and performance of the seed or seed lot during germination and seedling emergence."

AOSA:"Seed vigour comprises of those seed properties which determine the potential for rapid, uniform emergence and development of normal seedlings under a wide range of field conditions."

What causes vigour loss

Vigour loss is due mainly to seed deterioration and aging, which starts as soon as the seed becomes physiologically mature. It's imperative, therefore, that seed be handled carefully to prevent accelerated reduction in performance through physical damage to cell membranes. This is particularly true of large seeds, such as pulses and legumes, where seed damage is the primary cause of deterioration. Other associated detrimental factors include:

- Enzyme activity in immature and dormant seeds
- Respiration during harvest and storage
- Impaired protein and the use of protein and RNA synthesis during periods of low temperature stress
- Genetic damage
- Accumulation of toxic metabolites.

Methods of measuring seed vigour

The general strategy of determining seed vigour is to measure some aspects of seed deterioration or weaknesses, which is inversely proportional to seed vigour. Cold test, accelerated aging test, electric conductivity test, seedling vigour classification, and seedling growth rate are among the tests that are used to measure seed vigour. In addition, the tetrazolium (TZ test) can be used as a vigour test by classifying the pattern of stained seeds into high, medium and low quality.

Cold Test (CT)

The cold test simulates early spring field conditions by germinating the seeds in wet soils (>70% water holding capacity) and incubating them at 5-10°C/41-51°F for a specified period. At the end of the cold period, the test is transferred to a favourable temperature for germination (*e.g.*, 25°C/77°F in case of sweet corn). The percentage of normal seedlings is considered as an indication of seed vigour. Vigorous seeds germinate better under cold environments.

Accelerated Aging Test (AAT)

The principle of this test is to stress seeds with high temperatures of (40-45°C/ 130-139°F) and near 100% relative humidity (RH) for varying lengths of time, depending on the kind of seeds, after which a germination test is made. High vigour seeds are expected to tolerate high temperatures and humidity and retain their capability to produce normal seedlings in the germination test.

Electric Conductivity Test

This test measures the integrity of cell membranes, which is correlated with seed vigour. It is well established that this test is useful for garden beans and peas. It has been also reported that the conductivity test results are significantly correlated with field emergence for corn, and soybean. As seeds lose vigour, nutrients exude from their membranes, and so low quality seeds leak electrolytes such as amino acids, organic acids while high quality seeds contains their nutrients within well structured membranes. Therefore, seeds with higher conductivity measurement are indication of low quality seeds as vice versa.

Seedling Vigour Classification Test (SVCT)

This vigour test is an expansion of the standard germination test (SGT). The normal seedlings obtained from the SGT results are further classified into 'strong' and 'weak' categories. This test has been used for corn, garden beans, soybean, cotton, peanuts and other crops. Seedlings have four significant morphological sites for evaluating vigour: Root system, Hypocotyl (the embryonic axis between cotyledons and root), Cotyledons (storage tissue of reserve food for seedling development), Epicotyl (the embryonic axis above the cotyledons). In this test, seedlings are classified as 'strong' if the above four areas are well developed and free from defects, which is indication of satisfactory performance over a wide range of field conditions. On the other hand, normal seedlings with some deficiencies such as missing part of the root, one cotyledon missing, hypocotyl with breaks, lesions, necrosis, twisting, or curling are classified as 'weak'.

17

Seed Health Test

Seed Health Testing

Science of determining the presence or absence of disease causing agents such as fungi, bacteria and viruses and insects in the seed samples.

Seed Borne disease

The seed primordial or the maturing seed may be infected either (i) directly from the infected plant through the flower or fruit stalk and the seed stalk or directly from the seed surface, or (ii)infection from outside may be introduced through stigma or ovary wall or pericarp, and the flower or fruit stalk, and later through the seed coat. A pathogen may penetrate several of these parts of the seed and in turn infect them. The infestation /contamination of the seed may occur during harvesting, threshing and processing. Visualizing the importance of seed borne diseases the Central Seed Certification board has prescribed certification standards for foundation and certified seed for several diseases.

The pathogen may carry with the seeds in three ways.

- **Admixture:** Pathogens are independent of seeds but accompany them. They are mixed with health seeds during threshing.
- **External:**The pathogen may be present on seed surface as spores, oospores and chlamydospores.
- **Internal:** Pathogens establish within the seed with definite relationship with seed parts.

Effect of pathogens on seeds

1. Reduction in market value

a. Discolouration and Shrivelling

Seed borne fungi cause discolouration of part or whole of the seed. Due to discolouration and destoration market value of the seed is reduced.

b. Reduction in germination and vigour

Seeds after sowing start germinating and with the activity of germination seeds, seed borne, pathogens become active which may cause seed decay and / or pre or post emergence damping off. Eventually the plant stand is poor in the field. The reduction in germination depends upon the variety / cultivar, type amount and location of inoculum, environmental condition and other factors.

2. Disease Development

Seeds are the most important means of perpetuation of plant pathogens. Pathogens remain viable longer in seed than in vegetative plant parts or in soil. Transmission of a pathogen through seed is considered more important than other survival means the host parasite relationship within seeds also forms the earliest possible infection in the field. Since pathogens are in direct contact with the seeds, the chances of seedling infection are enhanced. The inoculum of the pathogen from seed may spread fast favourable conditions and cause an epidemic. On farms where disease plants contaminated seed are used as cattle feed and diseased plants are bedding, the spreading of manure can introduce the pathogen to uncontaminated fields.

A pathogen can be introduced to new areas through seeds. In early times the movement of pathogens was restricted to certain areas because of limited means of travel, but today the chances of spread through seeds to large areas have increased due to expanded and rapid modes of transport. In exchange of germplasm / plant the risk of introducing of new strain / physiologic races of a pathogen or new pathogen from one country other country has increased.

3. Biochemical changes in seeds

Various seed borne fungi brings about qualitative changes in the physico chemical properties of seeds such as colour, odour, oil content, iodine and saponification values, refractive index and protein content. These changes ultimately affect the commercial value of the seeds.

A. Protein

Certain fungi such *Aspergillus flavus, A niger*and *Rhizoctonia solani* in peanut seeds increased crude fat content but reduced starch content, probably due to amylase production by these fungi. *Macrophomina phaseolina* infection in peanut caused both qualitative and quantitative damage, including discolouration of pods and seeds and a reduction in pod and seed yield and oil content.

B. Oil

Oil is synthesized by condensation of molecules of glycerine or fatty acids or both. The glycerine and fatty acids are synthesized from carbohydrates, especially reducing sugars. Naturally occurring mixtures of glycerides in oil contain a small amount of free fatty acids.

Due to high lipase activity of fungi free fatty acids are increased. Wilson (1947) reported that microorganisms produced free fatty acids during the lipolysis of triglycerides. The chemical changes resulting in increased unsaturated fatty acids are undesirable and can cause fatness and cardiovascular disorders in humans. Hence, such seeds are not fit for oil extraction and human consumption or for derived oil cake as animal feed.

4. Changes in physical properties of seeds

Some important physical properties in seeds, *viz.*, density, shape, size, surface area, volume and weight were affected by pathogen infection.

Detection methods

Healthy and viable seed is the first pre-requisite for increasing seed production and to reduce possible seed crop failures. Until and unless we do not know the health status of the seed, it is not possible to manage the disease. To know health of the seed, different testing methods for different pathogens / diseases of different crops have been developed. The entire submitted sample, or a portion of it, depending on the test method, may be used. Normally the working sample shall not be less than 400 pure seeds.

Methods

1. Examination without incubation

Such tests give no indication as to the viability of the pathogen

a. Direct examination

The submitted sample, or a sub-sample from its is examined, with or without a stereoscopic microscopic and searched for ergots and other sclerotia, nematode galls, smut balls, insects, mites and evidence of diseases and pests in seed or in inert matter.

b. Examination of Imbibed seeds

The working sample is immersed in water or other liquid to make fruiting bodies symptoms of pests etc., more easily visible, or to encourage the liberation of spores. After imbibition the seeds are examined either superficially or internally preferably with the help of stereoscopic microscope.

c. Examination of Organisms removed by washing

The working sample is immersed in water with a wetting agent or alcohol and shaken vigorously to remove fungal spores, hyphae, nematodes, etc., intermingled with or adhering to the seeds. The excess liquid is then removed by filtration, centrifugation or evaporation and the extracted material examined with the help of compound microscope.

2. Examination after incubation

After incubation for a specific period, the working sample is examined for the presence of or symptoms of disease organisms, pests, and evidence of physiological disturbances in the seeds and seedlings. The examination may be superficial or internal. Three types of media are commonly used.

- Blotters, these are used when pathogens are to be grown from the seeds or when seedlings are to be examined. The seeds with or without pre treatment are so spaced during incubation as to avoid secondary spread of organisms. Lighting is provided to stimulate sporulation of fungi when needed. Some pathogens can be identified without magnification but a stereoscopic microscope or a compound microscope is often helpful in identifying spores.

- Sand, artificial composts and similar media can be used for certain pathogens. The seeds, usually without pre-treatment, are sown suitably spaced in the medium so as to avoid secondary spread of organisms and then incubated in conditions favourable for symptom expression.

- Agar plates are used to obtain identifiable growth of organisms form seeds. Precautions should be taken to ensure their sterilization. The seeds, normally after pre treatment are spaced on the surface of sterilized agar and incubated. Characteristic colonies on the agar can be identified, either macroscopically or microscopically. Lighting is often useful and germination inhibitors may be used.

3. Examination of plants

Growing plants from seed and examining them for disease symptoms is sometimes the most practicable method for determining whether bacteria, fungi or viruses are present in the sample. Seeds from the sample under test may be sown or inoculam obtained from the sample may be used for infection tests with healthy seedlings or parts of plants. The plants must be protected from accidental infections from elsewhere and conditions may require careful control.

4. Other techniques

Specialized methods involving serological reactions, phage-plaque formation, etc., have been developed for some disease organisms and may be used preferably in consultation with the seed pathologist.

Management of seed borne pathogens / diseases in seed quality programmes

1. Selection of seed production area
2. Eradication or reduction of soil borne pathogens
    - Crop rotation
    - Fallow
    - Burning
    - Water
3. Modification of cultural practices
    - Preparation of the seed bed
    - Date of sowing
    - Depth of sowing
    - Rate of sowing
    - Proper dose of fertilizers
    - Selection of disease resistant /tolerant varieties
4. Reduction or elimination of seed-borne inoculum
    - Certification and field inspection of seed crops
    - Biological control
    - Seed treatment
    - Isolation distance
    - Eradication of other hosts.

5. Chemical protection of seed crops
6. Management of storage fungi during storage
7. Plant quarantine

Seed borne insects

Insect pest infestation during storage leads to poor health status of seeds. The seed quality is affected on the basis of insect damage to endosperm or embryo. If embryo portion is damaged, seed fails to germinate and if endosperm is damaged, seed may germinate and develop into a seedling but the vigour of the seedling will depend upon the extent and intensity of the damage. The degree and type of insect infestation and insect species involved depend upon the crop seed and prevailing ecological conditions. Other factors *viz.*, sanitation, bulk or bag storage etc., generally influence of the seed safety. In bulk storage seeds continuously release heat and moisture by respiration, slowing down the air movement through the bulk of the seeds, which tends to develop hot spots and caked seeds. Insects also damage indirectly by contaminating the seed with their waste, caste skins, webbing and body parts. Heating also takes place due to feeding by some insect pests which leads to the development of moulds.

A large number of insect species are associated with seeds during storage. These insects mostly belong to orders Coleoptera (beetles / weevils) and Lepidoptera (moths). These insects, depending upon their ability to damage sound seeds can be divided into Primary and Secondary seed feeders. The primary consumers breach the pericarp or testa making feeding possible by a much wider range of insects, mites and fungi.

The primary consumer obtains water metabolically from seeds and releases it into their immediate environment along with waste heat. At the same time the accumulating frass, seed dust and carcasses restrict air movement, contributing to the formation of hot spots. Amongst primary pests, adults and larvae of *Trogodemagranarium, Sitophilus oryzae* and *Rhizopertha dominica* are most important pests. The generation time is often short and in a few months, the population can explode. The weevils complete their life cycle inside individual seeds and may develop into substantial population before their presence is even noticed.

Secondary seed feeders attack the already damaged seed where testa is cracked, holed and broken either due to mechanical damage during harvesting, threshing and processing or by too rapid seed drying or by prior feeding damage by primary seed feeders. Such insects are primary pests of flour / processed products. Most common secondary pests on seeds include the larvae of *Ephestia, plodia* and the beetles of *Oryzaephilus*, *Crytolests*, *Triboliumand* other *Tenebrionidae*. As per the climatic requirements, *Trogodermais* a serious pest of warm and dry regions while *Sitophilus* and *Rhizopertha* of humid and moderately warm regions.

In seed stores insects can be categorized in to two groups.

Group I. Insects that infests the seed in the field and do not multiply in transit / store *e.g.*, pink boll worm in cotton, midges in millets etc., The number of seeds damaged in field remains constant. Pests are transmitted through these infested seeds from place to place and year to year. They resume their life activity in field or other media under favourable conditions.

Group II. Insects that infest the seed in field / store or any other place where the seed is handled. The arthropod pests continue life-cycle in ambient conditions. Number of damaged seeds as well insects continue to increase with the passage of time *e.g.*, weevils, beetles, moths etc., Thus, after lifting a seed sample containing such insects, it should immediately be tested.

Detection of External Infestation : Direct Examination

a. Presence of living insects

Seeds are examined in dry state with the help of magnifying glass (10X) or stereoscopic microscope aided with light. Adult weevil, beetle, moth larvae, grubs etc. are separated and counted. The result is reported as number of insects (including its stages) per weight of the sample.

b. Detection of external injury

Seeds are thoroughly examined with the help of magnifying glass or stereoscopic microscope aided with light. Seeds damaged (less than one-half of its size) by insects including those seeds whose germ (embryo) has been touched or eaten, having escape holes, adhered with eggs etc. are separated and counted. Other seeds with no visible symptoms of insect injury are subjected to further test to detect internal infestation.

c. Detection of Internal (hidden) infestation

Some of the stored-seed pests especially weevils and beetles are internal feeder. Detection of their infestation is essential to determine actual infestation. The number of infested seeds, thus obtained is added to the number of seeds found externally damaged by the insects for final calculation.

Alkali or Glycerine Method

Seeds are taken in a beaker (100 ml capacity) and then Sodium hydroxide (NaOH) 10% solution (dissolve 10 gm NaOH pellets in 100 ml water) is poured in it until the seeds are submerged. After decanting solution, translucent seeds are washed with water and then examined with the help of magnifying glass. Seeds with visible internal infestation are separated, cut opened to confirm the presence of insect or its stage and counted. The result is reported after adding the count of damaged seeds obtained in tests as percentage (by number). Alternatively, the seeds can also be made translucent in lacto phenol solution (dissolve 20 gm phenol crystal in 20 ml luke warm distilled water, and then add 20 ml lactic acid and 10 ml glycerine) following the steps mentioned above.

Management

1. Preventive

- Prompt harvest, drying and thorough cleaning of the seed help prevent entry of insects into the seed stores.
- Good sanitation in the processing plant and harvest machinery is crucial, since seed pests, often breed in small remnant deposits of old seeds in accessible places *viz.*, conveyor belts, screens, lift hoppers etc.,
- Hard to reach spots should be disinfected by fumigation.
- Destroy (burn or bury deep) the sweepings, clear trash, litter from outside the store / processing hall and remove the spilled seeds from under the stacks.
- Remove all the leftover seeds from store/bin; sweep down the walls, ceilings, sills, ledges and floor etc.,
- Make necessary repairs of crack and crevices.
- Use new gunny bags for fresh harvest. However, if old bags are to be used dip them in 0.1% Malathion 50 EC (1 part Malathion 50 EC + 500 parts of water).

2. Curative

Under tropical and sub-tropical climate conditions, inadequate methods of seed storage as well as high risk factors of insect infestation warrants the use of fumigants in seed stores or application of insecticides to save the seed from insect ravages. Fumigation with fumigants is widely practices, for curative or preventive action, as they are cost effective, efficient and easy to use against target pests besides can penetrate into places when other control methods become impossible or impractical.

Fumigant is a chemical substance, with at specified temperature and pressure can exist in gaseous from in sufficient concentration to be lethal to the pest organisms. Apart from other characteristics a fumigant should not affect the viability of seeds, should be non-persistent, non-corrosive and highly diffusible. Fumigation with fumigants *viz.*, Aluminium phosphide, Methyl bromide, Ethyl di-bromide and EDCT are most common world over. Fumigate with Aluminium phosphate @ 7 tablets (3g each) or EDCT mixture (3:1) per 1000 cubic feet space with exposure period of at least 7 days. Never keep the fumigants at the bottom of the floor/bin as the gas is heavier than air and travel downwards. Since the gas can penetrate downwards unto 8 feet, the fumigant should be placed accordingly.

Spray the stores with DDVP @ 0.25% (1:300). For 100m^2 area 3 liters of spray material is required. Spray on all the walls, ceiling and floor of the store. Spray other surfaces/structures (see checklist) where presence of insect is suspected.Surface of the bags should be sprayed with Malathion 50 EC (1 ml Malathion in 100 ml water after 3 or weeks).

3. Non Chemical Methods

a. Biological control

Augmentation or manipulation of these natural enemies offers new means of controlling storage pests in situations where the use of other means of control is objectionable. A parasitic wasp, *Bracon hebetor* attacks late stage larvae of *Cautella* and *Plodiainterpunctella* and suppresses their populations. The potential of this method is still rudimentary in stores.

b. Pheromones

Synthetic pheromones have produced for the almond moth, red and confused flour beetles, lesser gain borer, *khapra* and few other insects species. The studies have shown that pheromone traps are effective in detecting hidden infestation.

c. Host resistance to insect-pest

Though much effort by plant breeders has been made to develop insect resistant lines in various crops, yet little research has been performed to develop lines whose grains / seeds resist attack by stored product insects.

d. Temperature manipulation

The application of high or low temperatures offers a non-chemical means of disinfecting stored commodities. Sub lethal temperatures can affect reproduction, growth, development, feeding and movement. However, temperature adverse to survival of storage insect population must not affect the seed viability.

e. Controlled or modified atmospheres

Killing of insects with modified atmosphere of oxygen, carbon di oxide and nitrogen has long been recognized. Laboratory studies have shown that atmospheres deficient in oxygen are effective in controlling all life stages of principal insect species that infest stores.

f. Radiation

Effect of Gamma-Radiation has been studies extensively for disinfestations of insects in stores. However, its affect on seed viability has not studied in detail.

g. Physical barriers

One of the demonstrated non-chemical methods for the control of stored pests is the use of physical barriers placed around the commodity. For example, a mutli-wall paper bag has been developed to protect the seeds. Another physical barrier is the packaging material, which prevents the insects from infesting the seeds in bags. The packaging material identified to be resistant to insects are; ethylene tetrafluro ethylene, polyester, polyvinylchloride, polypropylene and polycarbonate. Insecticides have also been inserted into packaging material to make them insect resistant. Plastic sheet with high tensile strength is also being tried as barrier for the insect infiltration. Also, physical disturbance or turning of seeds vacuums conveyance and Entoleters has been reported to reduce the population of external feeding insects only. Inert dusts derived from the shells of diatoms are being used to mix with seeds. However seed moisture must be kept in mind while using these dusts.

18

The Seeds Act, 1966

An Act to provide for regulating the quality of certain seeds for sale, and for matters connected therewith. It enacted by Parliament in the Seventeenth Year of the Republic of India.

1. Short Title, Extent and Commencement

- This Act may be called the Seeds Act, 1966.
- It extends to the whole of India.
- It shall come into force on such date as the Central Government may, by notification in the Official Gazette, appoint, and different dates may be appointed for different provisions of this Act, and for different States or for different areas thereof.

2. Definitions

In this Act, unless the context otherwise requires:

- Agriculture includes horticulture
- Central Seed Laboratory means the Central Seed Laboratory established or declared as such under sub-section (1) of section 4
- Certification agency means the certification agency established under Section 8 or recognised under Section 18
- Committee means the Central Seed Committee constituted under sub-section (1) of Section 3
- Container means a box, bottle, casket, tin, barrel, case, receptacle, sack, bag, wrapper or other thing in which any article or thing is placed or packed
- Export means taking out of India to a place outside India
- Import means bringing into India from a place outside India

- Kind means one or more related species or sub-species of crop plants each individually or collectively known by one common name such as cabbage, maize, paddy and wheat
- Notified kind or variety, in relation to any seed, means any kind or variety thereof notified under Section 5
- Prescribed means prescribed by rules made under this act
- Seed means any of the following classes of seeds used for sowing or planting, seeds of food crops including edible oil seeds and seeds of fruits and vegetables, cotton seeds, seeds of cattle fodder, and includes seedlings, and tubers, bulbs, rhizomes, roots, cuttings, all types of grafts and other vegetatively propagated material, of food crops or cattle fodder
- Seed Analyst means a Seed Analyst appointed under section 12
- Seed Inspector means a Seed Inspector appointed under section 13
- State Government, in relation to a Union territory, means the administrator thereof
- State Seed Laboratory, in relation to any State, means the State Seed Laboratory established or declared as such under sub-section (2) of section 4 for that State; and
- Variety means a sub-division of a kind identifiable by growth, yield, plant, fruit, seed, or other characteristic.

3. Central Seed Committee

(1) The Central Government shall, as soon as may be after the commencement of this Act, constitute a Committee called the Central Seed Committee to advise the Central Government and the State Governments on matters arising out of the administration of this Act and to carry out the other functions assigned to it by or under this Act.

(2) The Committee shall consist of the following members, namely, Chairman to be nominated by the Central Government; eight persons to be nominated by the Central Government to represent such interests that Government thinks fit, of whom not less than two persons shall be representatives of growers of seed; one person to be nominated by the Government of each of the States.

(3) The members of the Committee shall, unless their seats become vacant earlier by resignation, death or otherwise, be entitled to hold office for two years and shall be eligible for renomination.

(4) The Committee may, subject to the previous approval of the Central Government, make bye-laws fixing the quorum and regulating its own procedure and the conduct of all business to be transacted by it.

(5) The Committee may appoint one or more sub-committees, consisting wholly of members of the Committee or wholly of other persons or partly of members of the Committee and partly of other persons, as it thinks fit, for the purpose of discharging such of its functions as may be delegated to such sub-committee or sub-committees by the Committee.

(6) The functions of the Committee or any sub-committee thereof may be exercised notwithstanding any vacancy therein.

(7) The Central Government shall appoint a person to be the secretary of the Committee and shall provide the Committee with such clerical and other staff as the Central Government considers necessary.

4. Central Seed Laboratory and State Seed Laboratory

- The Central Government may, by notification in the Official Gazette, establish a Central Seed Laboratory or declare any seed laboratory as the Central Seed Laboratory to carry out the functions entrusted to the Central Seed Laboratory by or under this Act.
- The State Government may, by notification in the Official Gazette, establish one or more State Seed Laboratories or declare any seed laboratory as a State Seed Laboratory where analysis of seeds of any notified kind or variety shall be carried out by Seed Analysts under this Act in the prescribed manner.

5. Power to notify kinds or varieties of seeds

If the Central Government, after consultation with the Committee, is of opinion that it is necessary or expedient to regulate the quality of seed of any kind or variety to be sold for purposes of agriculture, it may, by notification in the Official Gazette, declare such kind or variety to be a notified kind or variety for the purposes of this Act and different kinds or varieties may be notified for different States or for different areas thereof.

6. Power to specify minimum limits of germination and purity, etc.

The Central Government may, after consultation with the Committee and by notification in the Official Gazette, specify the minimum limits of germination and purity with respect to any seed of any notified kind or variety; the mark or label to indicate that such seed conforms to the minimum limits of germination and purity specified under clause (a) and the particulars which such mark or label may contain.

7. Regulation of sale of seeds of notified kinds or varieties

No person shall, himself or by any other person on his behalf, carry on the business of selling, keeping for sale, offering to sell, bartering or otherwise supplying any seed of any notified kind or variety, unless such seed is identifiable as to

a. its kind or variety; such seed conforms to the minimum limits of germination and purity specified under clause (a) of section 6;

b. the container of such seed bears in the prescribed manner, the mark or label containing the correct particulars thereof, specified under clause (b) of section 6; and

c. he complies with such other requirements d. as may be prescribed.

8. Certification agency

The State Government or the Central Government in consultation with the State Government may, by notification in the Official Gazette, establish a certification agency for the State to carry out the functions entrusted to the certification agency by or under this Act.

9. Grant of certificate by certification agency

(1) Any person selling, keeping for sale, offering to sell, bartering or otherwise supplying any seed of any notified kind or variety may, if he desires to have such seed certified by the certification agency, apply to the certification agency for the grant of a certificate for the purpose.

(2) Every application under sub-section (1) shall be made in such form, shall contain such particulars and shall be accompanied by such fees as may be prescribed.

(3) On receipt of any such application for the grant of a certificate, the certification agency may, after such enquiry as it thinks fit and after satisfying itself that the seed to which the application relates conforms to the minimum limits of germination and purity specified for that seed under clause (a) of section 6, grant a certificate in such form and on such conditions as may be prescribed.

10. Revocation of certificate

If the certification agency is satisfied, either on a reference made to it in this behalf or otherwise, that the certificate granted by it under section 9 has been obtained by misrepresentation as to an essential fact; or the holder of the certificate has, without reasonable cause, failed to comply with the conditions subject to which the certificate has been granted or has contravened any of the provisions of this Act or the rules made there under; then, without prejudice to any other penalty to which the holder of the certificate may be liable under this Act, the certification agency may, after giving the holder of the certificate an opportunity of showing cause, revoke the certificate.

11. Appeal

(1) Any person aggrieved by a decision of a certification agency under section 9 or section 10, may, within thirty days from the date on which the decision is communicated to him and on payment of such fees as may be prescribed, prefer an appeal to such authority as may be specified by the State Government in this behalf: Provided that the appellate authority may entertain an appeal after the expiry of the said period of thirty days if it is satisfied that the appellate was prevented by sufficient cause from filing the appeal in time.

(2) On receipt of an appeal under sub-section (1), the appellate authority shall, after giving the appellant an opportunity of being heard, dispose of the appeal as expeditiously as possible.

(3) Every order of the appellate authority under this section shall be final.

12. Seed Analysts

The State Government may, by notification in the Official Gazette, appoint such persons as it thinks fit, having the prescribed qualifications, to be Seed Analysts and define the areas within which they shall exercise jurisdiction.

13. Seed Inspectors

(1) The State Government may, by notification in the Official Gazette, appoint such persons as it thinks fit, having the prescribed qualifications, to be Seed Inspectors and define the areas within which they shall exercise jurisdiction.

(2) Every Seed Inspector shall be deemed to be a public servant within the meaning of section 21 of the Indian Penal Code (45 of 1860) and shall be officially subordinate to such authority as the State Government may specify in this behalf.

14. Powers of Seed Inspector

(1) The Seed Inspector may take samples of any seed of any notified a. kind or variety from any person selling such seed; or any person who is in the course of conveying, delivering or preparing to deliver such seed to a purchaser or a consignee; or a purchaser or a consignee after delivery of such seed to him; send such sample for analysis to the Seed Analyst for the area within which such sample has been taken; enter and search at all reasonable times, with such assistance, if any, as he considers necessary, any place in which he has reason to believe that an offence under this Act has been or is being committed and order in writing the person in possession of any seed in respect of which the offence has been or is being committed, not to dispose of any stock of such seed for a specific period not exceeding thirty days or, unless the alleged offence is such that the defect may be removed by the possessor of the seed, seize the stock of such seed; examine any record, register, document or any other material object found in any place mentioned in clause (c) and seize the same if he has reason to believe that

it may furnish evidence of the commission of an offence punishable under this Act; and exercise such other powers as may be necessary for carrying out the purposes of this Act or any rule made there under.

(2) Where any sample of any seed of any notified kind or variety is taken under clause (a) of

sub-section (1), its cost, calculated at the rate at which such seed is usually sold to the public, shall be paid on demand to the person from whom it is taken.

(3) The power conferred by this section includes power to break-open any container in which any seed of any notified kind or variety may be contained or to break-open the door of any premises where any such seed may be kept for sale: Provided that the power to break-open the door shall be exercised only after the owner or any other person in occupation of the premises, if he is present therein, refuses to open the door on being called upon to do so.

(4) Where the Seed Inspector takes any action under clause (a) of sub-section (1), he shall, as far as possible, call not less than two persons to be present at the time when such action is taken and take their signatures on a memorandum to be prepared in the prescribed form and manner.

(5) The provisions of the Code of Criminal Procedure, 1898 (5 of 1898), shall, so far as may be, apply to any search or seizure under this section as they apply to any search or seizure made under the authority of a warrant issued under section 98 of the said Code.

15. Procedure to be followed by Seed Inspectors

(1) Whenever a Seed Inspector intends to take sample of any seed of any notified kind or variety for analysis, he shall give notice in writing, then and there, of such intention to the person from whom he intends to take sample; except in special cases provided by rules made under this Act, take three representative samples in the prescribed manner and mark and seal or fasten up each sample in such manner as its nature permits.

(2) When samples of any seed of any notified kind or variety are taken under sub-section (1), the Seed Inspector shall deliver one sample to the person from whom a. it has been taken; send in the prescribed manner another sample for analysis to the Seed Analyst for the area within which such sample has been taken; and retain the remaining sample in the prescribed manner for production in case any legal proceedings are taken or for analysis by the Central Seed Laboratory under sub-section (2) of section 16, as the case may be.

(3) If the person from whom the samples have been taken refuses to accept one of the samples, the Seed Inspector shall send intimation to the Seed Analyst of such refusal and thereupon the Seed Analyst receiving the sample for analysis shall divide it into two parts and shall seal or fasten up one of those parts and shall cause it, either upon receipt of the sample or when he delivers his report, to be delivered to the Seed Inspector who shall retain it for production in case legal proceedings are taken.

(4) Where a Seed Inspector takes any action under clause (c) of sub-section (1) of section 14: he shall use all despatch in ascertaining whether or not the seed contravenes any of the provisions of section 7 and if it is ascertained that the seed does not so contravene, forthwith revoke the order passed under the said clause or, as the case may be, take such action as may be necessary for the return of the stock of the seed seized; if he seizes the stock of the seed, he shall, as soon as may be, inform a magistrate and take his orders as to the custody thereof; without prejudice to the institution of any prosecution, if the alleged offence is such that the defect may be removed by the possessor of the seed, he shall, on being satisfied that the defect has been so removed, forthwith revoke the order passed under the said clause.

(5) Where as Seed Inspector seizes any record, register, document or any other material object under clause (d) of sub-section (1) of section 14, he shall, as soon as may be, inform a magistrate and take his orders as to the custody thereof.

16. Report of Seed Analyst

(1) The Seed Analyst shall, as soon as may be after the receipt of the sample under sub-section (2) of section 15, analyse the sample at the State Seed Laboratory and deliver, in such form as may be prescribed, one copy of the report of the result of the analysis to the Seed Inspector and another copy thereof to the person from whom the sample has been taken.

(2) After the institution of a prosecution under this Act, the accused vendor or the complainant may, on payment of the prescribed fee, make an application to the court for sending any of the samples mentioned in clause (a) or clause (c) of sub-section (2) of section 15 to the Central Seed Laboratory for its report and on receipt of the application, the court shall first ascertain that the mark and the seal or fastening as provided in clause (b) of sub-section (1) of section 15 are intact and may then despatch the sample under its own seal to the Central Seed Laboratory which shall thereupon send its report to the court in the prescribed form within one month from the date of receipt of the sample, specifying the result of the analysis.

(3) The report sent by the Central Seed Laboratory under sub-section (2) shall supersede the report given by the Seed Analyst under sub-section (1).

(4) Where the report sent by the Central Seed Laboratory under sub-section (2) is produced in any proceedings under Section 19, it shall not be necessary in such proceedings to produce any sample or part thereof taken for analysis.

17. Restriction on export and import of seeds of notified kinds or varieties

No person shall, for the purpose of sowing or planting by any person (including himself), export or import or cause to be exported or imported any seed of any notified kind or variety, unless it conforms to the minimum limits of germination and purity specified for that seed under clause (a) of section 6; and its container bears, in the prescribed manner, the mark or label with the b. correct particulars thereof specified for that seed under clause (b) of section 6.

18. Recognition of seed certification agencies of foreign countries

The Central Govt. may, on the recommendation of the Committee and by notification in the Official Gazette, recognise any seed certification agency established in any foreign country, for the purposes of this Act.

19. Penalty

If any person, contravenes any provision of this Act or any rule made there under; or prevents a Seed Inspector from taking sample under this Act; or prevents a Seed Inspector from exercising any other power conferred on him by or under this Act; he shall, on conviction, be punishable, for the first offence with fine which may extend to five hundred rupees, and in the event of such person having been previously convicted of an offence under this section, with imprisonment for a term which may extend to six months, or with fine which may extend to one thousand rupees, or with both.

20. Forfeiture of property

When any person has been convicted under this Act for the contravention of any of the provisions of this Act or the rules made thereunder, the seed in respect of which the contravention has been committed may be forfeited to the Government.

21. Offences by companies

(1) Where an offence under this Act has been committed by a company, every person who at the time the offence was committed was in charge of, and was responsible to the company for the conduct of the business of the company, as well as the company, shall be deemed to be guilty of the offence and shall be liable to be proceeded against and punished accordingly: Provided that nothing contained in this sub-section shall render any such person liable to any punishment under this Act if he proves that the offence was committed without his knowledge and that he exercised all due diligence to prevent the commission of such offence.

(2) Notwithstanding anything contained in sub-section (1), where an offence under this Act has been committed by a company and it is proved that the offence has been committed with the consent or connivance of, or is attributable to any neglect on the part of, any director, manager, secretary or other officer of the company, such director, manager, secretary or other officer shall also be deemed to be guilty of that offence and shall be liable to be proceeded against and punished accordingly.

Explanation. – For the purpose of this section,- "company" means anybody corporate and includes a firm or other association of individuals; and "director", in relation to a firm, means a partner b. in the firm.

22. Protection of action taken in good faith

No suit, prosecution or other legal proceeding shall lie against the Government or any officer of the Government for anything which is in good faith done or intended to be done under this Act.

23. Power to give directions

The Central Government may give such directions to any State Government as may appear to the Central Government to be necessary for carrying into execution in the State any of the provisions of this Act or of any rule made thereunder.

24. Exemption

Nothing in this Act shall apply to any seed of any notified kind or variety grown by a person and sold or delivered by him on his own premises direct to another person for being used by that person for the purpose of sowing or planting.

25. Power to make rules

(1) The Central Government may, by notification in the Official Gazette, make rules to carry out the purpose of this Act.

(2) In particular and without prejudice to the generality of the fore-going power, such rules may provide, for the functions of the Committee and the travelling and daily allowances payable to members of the Committee and members of any sub-committee appointed under

sub-section(5) of section 3; the functions of the Central Seed Laboratory; the functions of a certification agency; the manner of marking or labelling the container of seed of any notified kind or variety under clause (c) of Section 7 and under clause (b) of section 17; the requirements which may be complied with by a person carrying on the business referred to in section 7; the form of application for the grant of a certificate under section 9, the particulars

it may contain, the fees which should accompany it, the form of the certificate and the conditions subject to which the certificate may be granted; the form and manner in which and the fee on payment of which an appeal may be preferred under section 11 and the procedure to be followed by the appellate authority in disposing of the appeal; the qualifications and duties of Seed Analysts and Seed Inspectors; the manner in which samples may be taken by the Seed Inspector, the procedure for sending such samples to the Seed Analyst or the Central Seed Laboratory and the manner of analysing such samples; the form of report of the result of the analysis under sub-section (1) or sub-section (2) of section 16 and the fees payable in respect of such report under the said sub-section (2); the records to be maintained by a person carrying on the business referred to in section 7 and the particulars which such records shall contain; and any other matter which is to be or may be prescribed.

(3) Every rule made under this Act shall be laid as soon as may be after it is made, before each House of Parliament while it is in session for a total period of thirty days which may be comprised in one session or in two successive sessions, and if, before the expiry of the session in which it is so laid or the session immediately following, both Houses agree in making any modification in the rule or both Houses agree that the rule should not be made, that rule shall, thereafter have effect only in such modified form or be of no effect, as the case may be; so however, that any such modification or annulment shall be without prejudice to the validity of anything previously done under that rule.

19

The Seeds Rules, 1968

Part I - Preliminary

1. Short title: These rules may be called the Seeds Rules, 1968

2. Definitions: In these rules, unless the contest otherwise requires.

a. "Act" means the Seeds Act, 1966

b. "Advertisement" means a. representations other than those on the label, disseminated in any manner or by any means relating to seed for the purposes of the act;

c. "Certification sample" means a sample of seed drawn by a certification agency or by a duly authorized representative of a certification agency established under section 3 or recognized under section 18 of the Act;

d. "Certification tag" means a tag or label of certain design to be specified by the certification agency and shall constitute the certificate granted by the certification agency;

e. "Certified seed" means seed that fulfills all requirements for certification provided by the Act and these rules and to the container of which the certification tag is attached.

f. "Certified seed producer" means a person who grows or distributes certified seed in accordance with the procedure and standards of the certification agency;

g. "Complete record" means the information which relates to the origin, variety, kind germination and purity of seed of any notified kind or variety offered for sale, sold or otherwise supplied:

h. "Form" means a form appended to these rules:

i. "Origin" means the State, Union, Territory or foreign country where the seed is grown and in case seeds of different origin are blended the label shall show the percentage of seed of each origin;

j. Processing" means cleaning, drying, treating, grading and other operations which would change the purity and germination of the seed and thus requiring re-testing to determine the quality of the seed, but does not include operations such as packaging and labelling"

k. "Section" means a section of the Act'

l. "Service sample" means a sample submitted to the Central Seed Laboratory or to a State Seed Laboratory for testing, the results to be used as information for seeding selling or labelling purposes;

m. "Treated" means that the seed has been subjected to an applications of a substance or process in such a manner as to reduce, control or repel certain disease organisms, insects, or any other pests attacking such seeds or seedlings growing therefore and for other purposes.

PART II - Central Seed Committee

3. Functions of the Central Seed Committee: In addition to the functions entrusted to the committee by the Act, the committee shall :

a. Recommend the rate of fees to be levied for analysis of samples by the Central and States Seed Testing Laboratories and for certification by the certification agencies;

b. Advice the Central or State Governments on the suitability of seed testing laboratories;

c. Send its recommendations and other concerning records to the Central Government;

d. Recommend the procedure and standards for certification, tests and analysis of seeds; and

e. Carry out such other functions as are supplemental, incidental or consequential to any of the functions conferred by the Act or these rules.

4. Travelling and daily allowances payable to Members of the Committee and its Sub-committees

The members of the committee and its sub-committee shall be entitled to draw travelling and daily allowances as specified below when they are called upon to attend a meeting of the committee or a sub-committee thereof:

a. An official member of the committee or its sub-committee shall be entitle to draw travelling and daily allowances in accordance with the rules of the Government under which he is for the time being employed and from the same source from which his pay and allowances are drawn.

b. A non official member shall be allowed travelling and daily allowances in accordance with the general orders issued in this behalf by the Central Government from time to time.

PART III - Central Seed Laboratory

5. Functions: In addition to the functions entrusted to the /central Seed Laboratory by the Act, the Laboratory shall carry out the following functions, namely;

a. Initiate testing programmes in collaboration with the State Seed Laboratories designed to promote uniformity in test results between all seed laboratories in India;

b. Collect data continually on the quality of seeds found in the market and make this data available to the Committee; and

c. Carry out such other functions as may be assigned to it by the Central Government from time to time.

PART IV- Seed Certification Agency

6. Functions of the certification agency: In addition to the functions entrusted to the certification agency by the Act, the Agency shall certified seeds of any notified kinds or varieties;

a. Certify seeds of any notified kinds or variety;

b. Outline the procedure of submission of applications and for growing, harvesting, processing, storage and labeling of seeds intended for certification till the end to ensure that seed lots finally approved for certification are true to variety and meet prescribed standards for certification under the Act or these rules;

c. Maintain a list of recognized breeders of seeds;

d. Verify, upon receipt of an application, for certification, that the variety is eligible for certification that the seed source need for planting was authenticated and the record of purchase is in accordance with these rules and the fees has been paid;

e. Take sample and inspect seed lots produced under the procedures laid down by the certification agency and have such samples tested to ensure that the seed conforms to the prescribed standards of certification;

f. Inspect seed processing plants to see that the admixtures of other kinds and varieties are not introduced.

g. Ensure that action at all stages *e.g.*, field inspection, seed processing, plant inspection, analysis of samples taken and issue of certificates (including tags, marks, labels and seals) is taken expeditiously;

h. Carry out educational programmes designed to promote the use of certified seed including a publication listing certified seed growers and sources of certified seed;

i. Grant certificate (including tags, marks, labels and seals etc.) in accordance with the provisions of the Act and these Rules;

j. Maintain such records as may be necessary to verify that seed plants for the production of certified seed were eligible for such planting under these rules;

k. Inspect fields to ensure that the minimum standards for isolation, rouging (where applicable) use of male sterility (where applicable) and similar factors are maintained at all times, as well as ensure that seed borne diseases are not present in the field to a greater than those provided in the standards for certification.

PART V - Marking or labelling

7. Responsibility for marking or labelling: When seed of a notified kind or variety is offered for sale under section 7 each container shall be marked or labelled in the manner herein after specified. The person whose name appears on the mark or label shall be responsible for the accuracy of the information required to appear on the mark or label so long as seed is contained in the unopened original container. Provided, however, that such person shall not be responsible for the accuracy of the statement appearing on the mark or label if the seed is removed from the original unopened container, or he shall not be responsible for the accuracy of the germination statement beyond the date of validity indicated on the mark or label.

8. Contents of the mark or label: There shall be specified on every mark or label :

i. Particulars, as specified by the Central Government under Clause (b) of section 6 of the act.

ii. A correct statement of the net content in terms of weight and expressed in metric system.

iii. Date of testing.

iv. If the seed in container has been treated.

a. Statement indicating that the seed has been treated.

b. The commonly accepted chemical of abbreviated chemical (generic) name of the applied substance; and

c. If the substance of the chemical used for treatment, and present with the seed is harmful to human beings or other vertebrate animals, a caution statement such as "Do not use for food, feed or oil purposes". The caution for mercurial and similarly toxic substance shall be the word "Poison" which shall be in type size, prominently displayed on the label in red.

v. The name and address of the person who offers for sale, sells or otherwise supplies the seed and who is responsible for its quality;

vi. The name of the seed as notified under section 5 of the act.

9. Manner of marking or labeling the container under clause (c) of section 7 and clause (b) of section 17 :

a. The mark or label containing the particulars of the seed as specified under clause (b) of section 6 shall appear on each container of seed or on a tag or mark or label attached to the container in a conspicuous place on the inner most container in which the seed is packed and on every other covering in which that container is packed and shall be legible.

b. Any transparent cover or any wrapper, case or other covering used solely for the purpose of packing of transport or delivery need not be marked or labeled.

c. Where by a provision of these rules, any particulars are required to be displayed on a label on the container, such particulars may, instead of being displayed on a label be attached, painted or otherwise indelibly marked on the container.

10. Mark or label not to contain false or misleading statement

The mark or label shall not contain any statement, claim, design, device, fancy name or abbreviation which is false or misleading in any particular concerning the seed contained in the container.

11. Mark or label not to contain reference to the act or rules contradictory to required particulars

The mark or label shall not contain any reference to the Act, or any of these rules or any comment on, or reference to, or explanation of any particulars or declaration required by the Act or any of these rules which directly or by implication contradicts, qualifies or modifies such particulars or declaration.

12. Denial of responsibility for mark or label content prohibited

Nothing shall appear on the mark or label or in any advertisement pertaining to any seed of any notified kind or variety which shall deny responsibility for the statement required by or under the Act to appear on such mark, label or advertisement.

PART VI- Requirements

13. Requirements to be complied with by a person carrying on the business referred to in section 7

1. No person shall sell, keep for sale, offer to sell barter or otherwise supply any seed of any notified kind or variety, after the date recorded on the container, mark or label as the date up to which the seed may be expected to retain the germination not less than that prescribed under clause (a) of section 6 of the Act.

2. No person shall alter, obliterate or deface any mark or label attached to the container of any seed.

3. Every person selling, keeping for sale, offering to sell, bartering or otherwise supplying any seed of notified kind or variety under Section 7, shall keep over a period of three years a complete record of each lots of seed sold except that any seed sample may be discarded one year after the entire lot represented by such sample has been disposed off. The sample of seed kept as part of the complete record shall be as large as the size notified in the official Gazette.

This sample, if required to be tested, shall be tested only for determining the purity.

14. Classes and sources of certified seed

1. There shall be three classes of certified seed, namely, foundation, registered and certified, and each class shall meet the following standards for that class.

   a. Foundation seed shall be the progeny of breeder's seed, or be produced from foundation seed which can be clearly traced to breeder's seed Production shall be supervised and approved by a seed certification agency and be so handled as to maintain specific genetic purity and identity and shall be required to meet certification standards for the crop being certified.

   b. Registered seed shall be the progeny of foundation seed that is so handled as to maintain its genetic identity and purity according to standard specified for the particular crop being certified.

c. Certified seed shall be the progeny of registered or foundation seed that is so handled as to maintain genetic identity and purity according to standards specified for the particular crop being certified.

2. At the discretion of the certification agency (when considered necessary to maintain adequate seed supplies) certified seed may be progeny of certified seed provided this reproduction may not exceed three generations and provided further that it is determined by the seed certification agency that the genetic purity will not be significantly altered.

PART VII - Certification of Seeds

15. Application for the grant of a certificate

Every application for the grant of a certificate under Sub-section (1) of section 9 shall be made in Form in accordance with the procedure outlined by the certification agency for submission of applications and contain the following particulars, namely:

a. the name, profession and place of residence of the applicant;

b. the name of the seed to be certified, its notified kind or variety;

c. class of the seed;

d. source of the seed;

e. limits of germination and purity of the seed;

f. mark or label of the seed

16. Fees

Every application under sub-section (1) of the section 9 shall be accompanied by a fee of Rs, 25 in cash.

17. Certificate

Every certificate granted under Sub Section 3) of sections 9 shall be in Form II and shall be granted by the certifications agency after making enquiries and satisfying itself in accordance with the provisions of the said sub section of the following conditions for the period to the specified by the certification agency namely :

I. The person to whom the certificate is granted under sub sections (3) of section 9 shall attach a certification tag to every container of the certified seed and shall follow the provisions in respect of marking or labeling provided by or under the act.

II. The certification tag shall contain the following particulars namely.

a. Name and address of the certification agency.
b. Kind and variety of the seed.
c. Lot no or other mark of the seed.
d. Name and address of the certified seed producer.
e. Date of issue of the certificate and of its validity.
f. An appropriate sign to designate certified seed.
g. An appropriate word denoting the class designation of the seed.

III. The colour of the certification tag shall be white for foundation seed purple for registered seed and blue for certified seed.

IV. The container of the certified seed shall carry a seal of such material and in form as the certification agency may determine and no container carrying a certification tag shall be sold by the person if the tag or seal has either been tampered with or removed.

V. The certification tag on the container shall specify.

a. The period during which the seed shall be used for sowing or planting.
b. That the use of seed after the expiry of the validity period by any person is entirely at his risk and the holder of the certificate shall not be responsible for any damage to the buyer of the seed.
c. That no one should purchase the seed if the seal or the certification tag has been tampered with.

VI. The holder of the certificate shall keep record of the details of each lot of the seed which is issued for sale in such form as to be available for inspection and to be easily identified by reference to the number of the lot as shown in the certification tag of each container and such records shall be retained in the case of a seed for which expiry date is fixed for a period of two years form the expiry of such date.

VII. The holder of the certificate shall allow any Seed Inspector, authorized in writing by the certification agency in that behalf, to enter with or without prior notice, the premises where the seeds are grown, processed and sold and to inspect premises, plant and the process of processing at all reasonable hours.

VIII. The holder of the certificate shall allow the Seed Inspector, authorized in writing by the certification agency, to inspect all registers and records maintained under these rules and to take samples of the seeds and shall supply to the Seed

Inspector such information as he may require for the purposes of ascertaining whether the conditions subject tom which the certificate has been granted, have been complied with.

IX. The holder of the certificate shall on request furnish to the certification agency from every lot of the seed or from such lot or lots as the said agency may from time to time specify, a sample of such quantity as the agency may consider adequate for any examination required to be made.

X. If the certificati0on agency so directs, the holder of the certificate shall not sell or offer for sale any lot in respect of which a sample is furnished under the proceeding clause until the agency authorizes the sale of such lot.

XI. The holder the certificate shall, on being directed by the certification agency that any part of a lot has been found by the said agency not to conform to prescribed standards of quality or purity specified by or under the Act, withdraw the remainder of that lot from sale and so far as may, in the particular circumstances of the case, be practicable, recall all issues already made from that lot.

XII. The holder of the certificate shall comply with the provisions of the Act and these Rules and with the directions given after not less than one month's notice by the certification agency to such holder.

The Certification agency shall, before granting the certificate, ensure that the Seed conforms to the standards laid down in the Manual known as "Indian Minimum Seed Certification Standards" published by the Central Seed Committee, as amended from time to time". (Amendment No.18-48/81-SD, dated 10th June, 1981).

PART VIII- APPEALS

18. The form and manner in which and the fee on payment of which the Appeal may be preferred :

1. Every memorandum of appeal under sub-section (1) of section 11 shall be in writing and shall be accompanied by a copy of the decision of the certification agency against which it has been preferred and shall set forth concisely and under distinct heads the grounds of objections to such decision without any argument, or narrative.

2. Every such memorandum of Appeal shall be accompanied by a treasure receipt for sum of 100/- rupees.

3. Every such memorandum of appeal may be presented either in person or through an agent duly authorized in writing in this behalf by the appellant or may be sent by the registered post.

19. Procedure to be followed by the appellate authority

In deciding appeals under the Act the Appellate authority shall exercise all the powers which a Court has and shall follow the same procedure with a Court follows in deciding appeals from the decree or order of an original Court under the Code of Civil Procedure, 1908 (5 of 1908).

PART IX - Seed Analysts and Seed Inspectors

20. Qualifications of Seed Analysts: A person shall not be qualified for appointment as Seed Analyst unless he.

i. Possesses a Master's or equivalent degree in Agriculture or Agronomy or Botany or Horticulture of a University recognized for this purpose by the Government and has had not less than one year's experience in seed technology; or

ii. Possessed a Bachelor's degree in Agriculture or Botany of a University recognized for this purpose by the Government and has had not less than three year's experience in seed technology.

21. Duties of a seed analyst

1. On receipt of a sample for a analysis the Seed Analyst shall first ascertain that the mark and the seal or fastening as provided in clause (b) of the subsection (1) of section 15 are intact and shall not the condition of the seals thereon.

2. The Seed analyst shall analyze the samples in accordance with the procedures laid down in the Seed Testing Manual published by the Indian Council of Agriculture Research as amended from time to time. (Amendment No.7(17)/69-Seeds-Dev., dated 30-6-1973)

3. The Seed Analyst shall deliver in Form VII, a copy of report of the result of the analysis to the persons specified in sub-section (1) of section 16, as soon as ,may be but not later than 30 days from the date of receipt of samples sent by the Seed Inspector under Sub-Section (2) of the Section 15.(Amendment No.7(17)/69-Seeds-Dev., dated 30-6-1973)

4. The Seed Analyst shall from time to time forward to the State Government the reports giving the result of analytical work done by him.

22. Qualification of Seed Inspectors

A person shall not be qualified for appointment as Seed Inspector unless he is a graduate in Agriculture of a University recognized for the purpose by the Government and has had not less than one year's experience in seed production, or seed development in seed analysis or testing in seed testing laboratory.

23. Duties of a Seed Inspector

In addition to the duties specified by the Act, the Seed Inspector shall

a. Inspect as frequently as may be required by certification agency all places used for growing storage or sale of any seed of any notified kind or variety;

b. Satisfy himself that the conditions of the certificates are being observed;

c. Procedure and send for analysis, if necessary, samples of any seeds, which he has reason to suspect are being produced, stocked or sold or exhibited for sale on contravention of the provisions of the Act or these rules;

d. Investigate any complaint, which may be made to him in writing in respect of any contravention of the provisions of the Act or these rules.

e. Maintain a record of all inspections made and action taken by him in the performance of his duties including the taking of samples and the seizure of stocks and submit copies of such record to the Director of Agriculture or the certification agency as may be directed in this behalf.

f. When so authorized by the State Government detain imported containers which he has reason to suspect contain seeds, import of which is prohibited except and in accordance with the provisions of the Act or these rules.

g. Institute prosecutions in respect of breaches of the Act or these rules.

h. Perform such other duties as may be entrusted to him by the State Government (Amendment No.7(17)/69-Seeds-Dev., dated 30-6-1973)

23 (A) Action to be taken by the Seed Inspector if a complaint is lodged with him

1. If farmer has lodged a complaint in writing that the failure of the crop is due to the defective quality of seeds of any notified kind or variety supplied to him, the Seed Inspector shall take in his possession the marks or labels, the seed containers and a sample of unused seeds to the extent possible from the complaint for establishing the sources of supply of seeds and shall investigate the causes of the failure of his crop by sending samples of the lot to the Seed Analyst for detailed analysis at the State Seed Testing Laboratory. He shall thereupon submit the report of his findings as soon as possible to the competent authority.

2. In case, the Seed Inspector comes to the conclusion that the failure of the crop is due to the quality of seeds supplied to the farmer being less than the minimum standards notified by the Central Government, he shall launch proceedings (Amendment No.7-15/74-SD, Dated 31 January, 1976) against the supplier for contravention of the provisions of the Act or these Rules". (Amendment No.7-15/74.SD, dated 29th April, 1975)

PART X - Sealing, fastening, dispatch and analysis of samples

24. Manner of taking samples

Samples of any seed of any notified kind of variety for the purpose of analysis shall be taken in a clean dry container which shall be closed sufficiently tight to prevent leakage and entrance of moisture and shall be carefully sealed.

25. Containers to be labeled and addressed

All containers containing samples for analysis shall be properly labeled and the parcels shall be properly addressed. The label on any sample of seed sent for analysis shall bear.

a. Serial number;

b. Name of the sender with official designation, if any;

c. Name of the person from whom the sample has been taken.

d. Date and place of taking the sample;

e. Kind or variety of the seed for analysis;

f. Nature and quantity of preservative, if any, added to the sample.

26. Manner of packing, fastening and sealing the samples

All samples of seed sent for analysis shall be packed, fastened and sealed in the following manner.

a. The stopper shall first be securely fastened to as to prevent leakage of the containers in transit;

b. The container shall then be completely wrapped in fairly strong thick paper. The ends of the paper shall be neatly folded in and affixed by means of gum or other adhesive.

c. The paper cover shall be further secured by means of strong twine or thread both above and across the container, and the twine or thread shall then be fastened on the paper cover by means of sealing wax on which there shall be test four distinct and clear impression of the seal of the sender of which one shall be at the top of the packet, one at the bottom and the other two on the body of the packet. The knots of the twine or thread shall be covered by means of sealing wax bearing the impression of the seal of the sender.

27. Form of order

The order to be given in writing by the Seed Inspector under clause (c) of subsection (1) of section 14 shall be Form III

28. Form of receipt of records

When a Seed Inspector seizes any record, register, document or any other material object under Clause (d) of Sub-section (1) of section 14, he shall issue a receipt in Form IV to the person concerned.

29. Samples how to sent to the Seed Analyst

The container of sample for analysis shall be sent to the Seed Analyst by registered post or by hand in a sealed packet enclosed together with a memorandum in Form V in an outer cover addressed to the Seed Analyst.

30. Memorandum and impression of seal to be sent separately

A copy of the memorandum and a specimen impression of the seal used to seal the packet shall be sent to the Seed Analyst separately by registered post or delivered to him or to any person authorized by him.

31. Addition of preservatives to samples

Any person taking a sample of seed for the purpose of analysis under the Act may add a preservative as may be specified from time to time to the sample the nature and quantity of the preservative added shall be clearly noted on the label to be affixed to the container.

32. Nature and quantity of the preservative to be noted on the label

Whenever any preservative is added to a sample the nature and quantity of the preservative added shall be clearly noted on the label to be affixed to the container.

33. Analysis of the sample

On receipt of the packet, it shall be opened either by the Seed Analyst or by an officer authorized in writing in that behalf by the Seed Analyst, who shall record the condition of the seal on the packet. Analysis of the sample shall be carried out at the State Seed Laboratory in accordance with the procedure laid down by the Central Government.

34. Form of notice

The notice to be given under clause (a) of sub section (1) of section 15 to the person from whom the Seed Inspector intends to take sample shall be in Form VI.

35. Form of report

The report of the result of the analysis under subsection (1) or subsection (2) of section 16 shall be delivered or sent in Form VII.

36. Fees

The fees payable in respect of the report from the Central Seed Laboratory under sub-section (2) of the 16 shall be Rs.10/- per sample of the seed analyzed.

37. Retaining of the sample

The sample of any seed shall, under clause (c) of Sub-section (2) of section 15, be retained under a cool, dry environment to eliminate the loss of viability and in insect proof or rat proof containers. The containers shall be dusted with suitable insecticides and the storage room fumigated to avoid infestation of samples by insects. The samples shall be packed in good quality containers of uniform shape and size before storage.

PART XI - Miscellaneous

38. Records: A person carrying on the business referred to in section 7 shall maintain the following records, namely:

a. Stock record of seed

b. Record of the sale of seeds

39. Form of Memorandum: The memorandum to be prepared under sub-section (4) of section 14 shall be in Form VIII.

20

The Seeds (Control) Order, 1983

Preliminary

1. Short title and extent

(i) This Order may be called the Seeds (Control) Order, 1983

(ii) It extends to the whole of India

(iii) It shall come into force on the 30th December, 1983

2. Definitions

In this order, unless the context otherwise requires,

(a) "Act" means the Essential Commodities Act, 1955 (10 of 1955)

(b) "Controller" means a person appointed as Controller of Seeds by the Central Government and includes any person empowered by the Central Government to exercise all or any functions of the Controller under this Order

(c) "Dealer" means a person carrying on the business of selling, exporting or importing seeds, and includes an agent of a dealer

(d) "Export" means to take or cause to be taken out from any place in India to a place outside India

(e) "Form" means a form appended to this Order

(f) "Import" means to bring or cause to be brought to any place in India from outside India

(g) "Inspector" means an inspector of seeds appointed under clause 12

(h) "Registering authority" means a licensing authority appointed under clause 11

(i) "Seeds" means the seeds as defined in the Seeds Act, 1966 (54 of 1966)

(j) "State Government" in relation to a Union Territory means the Administrator thereof by whatever designation known.

Dealer In Seeds to be Licensed

3. Dealer to obtain licence

(1) No person shall carry on the business of selling, exporting or importing seeds at any place except under and in accordance with the terms and conditions of licence granted to him under this order.

(2) Notwithstanding anything contained in sub-clause (1), the State Government may, by notification in the Official Gazette, exempt from the provisions of that sub-clause such class of dealers in such areas and subject to such conditions as may be specified in the notification.

4. Application for licence

Every person desiring to obtain a licence for selling, exporting or importing seeds shall make an application in duplicate in Form 'A' together with a fee of rupees fifty for licence to licensing authority.

5. Grant and refusal of licence

(1) The licensing authority may, after making such enquiry as it thinks fit, grant a licence in Form 'B' to any person who applies for it under clause 4. Provided that a licence shall not be issued to a person

(a) Whose earlier licence granted under this Order is under suspension, during the period of such suspension

(b) Whose earlier licence granted under this Order has been cancelled, within a period of one year from the date of such cancellation

(c) Who has been convicted under the Essential Commodities Act, 1955 (10 of 1955) or any order issued there under within three years preceding the date of application

(2) When the licensing authority refuses to grant licence to a person who applies for it under clause 4, he shall record his reasons for doing so.

6. Period of validity of licence

Every licence under this Order, shall, unless previously suspended or cancelled, remain valid for three years from the date of its issue.

7. Renewal of licence

(1) Every holder of licence desiring to renew the licence, shall, before the date of expiry of the licence, make an application for renewal in duplicate, to the licensing authority in Form 'C' together with a fee of rupees twenty for renewal. On receipt of such application, together with such fee, the licensing authority may renew the licence.

(2) If any application for renewal is not made before the expiry of the licence, but is made within one month from the date of expiry of the licence, the licence may be renewed on payment of additional fee of rupees twenty five, in addition to the fee for renewal of licence.

8. Dealers to display stock and price list

Every dealer of seeds shall display in his place of business

(a) The opening and closing stocks, on daily basis, of different seeds held by him

(b) A list indicating prices or rates of different seeds.

9. Dealers to give memorandum to purchaser

Every dealer shall give a cash or credit memorandum to a purchaser of seeds.

10. Power to distribute seeds

Where it is considered necessary to do so in public interest, the Controller may, by an order in writing direct any producer or dealer to sell or distribute any seed in such manner as may be specified therein.

Enforcement Authority

11. Appointment of licensing authority

The State Government may by notification in the Official Gazette appoint such number of persons as it thinks necessary to be licensing authority and may also define in that notification the area within which each such licensing authority shall exercise his jurisdiction.

12. Appointment of inspectors

The State Government may by notification in the Official Gazette appoint such number of persons as it thinks necessary to be inspectors and may in such notification define the local area within which each such Inspector shall exercise his jurisdiction.

13. Inspection and punishment

(1) An Inspector may with a view to securing compliance with this Order

(a) Require any dealer to give any information in his possession with respect to purchase, storage and sale of seeds by him

(b) Enter upon and search any premises where any seed is stored or exhibited for sale to ensure compliance with the provisions of this order

(c) draw samples of seeds meant for sale, export and seeds imported, and send the same in accordance with the procedure laid down in Schedule I, to a laboratory notified under the Seeds Act, 1966 (54 of 1966) to ensure that the sample conforms to standard of quality claimed

(d) Seize or detain any seed in respect of which he has reason to believe that a contravention of this Order has been committed or is being committed

(e) Seize any books of accounts or document relating to any seed in respect of which he has reason to believe that a contravention of this order has been committed or is being committed. Provided that the Inspector shall give a receipt, in respect of the books of accounts or documents seized, to the person from whom they have been seized. Provided further that the seized books of accounts or documents shall be returned to the person from whom the same had been seized after copies thereof or extracts there from as certified by such person have been taken.

(2) Subject to the provision of paragraph (d) of sub-clause (1), the provision, of section 100 of the Code of Criminal Procedure, 1973 (2 of 1974) relating to search and seizure shall, so far as may be, apply to searches and seizures under this clause.

(3) Where any seed is seized by an Inspector under this clause, he shall forthwith report the fact of such seizure to a Magistrate where-upon the provisions of sections 457 and 458 of the Code of Criminal Procedure, 1973 (2 of 1974) shall, so far as may be, apply to the custody and disposal of such seed.

(4) Every person, if so required by an Inspector, shall be bound to offer all necessary facilities to him for the purpose of enabling him to exercise his power under this clause.

14. Time limit for analysis

The laboratory to which a sample has been sent by an Inspector for analysis under this :

Order shall analyse the said samples and send the analysis report to the concerned Inspector within 60 days from the date of receipt of the sample in the laboratory.

15. Suspension/Cancellation of licence

The licensing authority may, after giving the holder of the licence an opportunity of being heard, suspend or cancel the licence on the following grounds, namely:

(a) That the licence had been obtained by misrepresentation as to a material particular; or

(b) That any of the provisions of this Order or any condition of licence has been contravened.

16. Appeal

Any person aggrieved by an order

(a) Refusing to grant, amend or renew the licence for sale, export or import of seeds

(b) Suspending or cancelling any licence, may within sixty days from the date of the order, appeal of such authority as the State Government may specify in this behalf, and the decision of such authority shall be final.

Provided that an application for appeal shall accompany an appeal fee of rupees fifty.

Miscellaneous

17. Amendment of licence

The licensing authority may, on receipt of a request in writing together with a fee of rupees ten from a dealer, amend the licence of such dealer.

18. Maintenance of records and submission of returns, etc.

(1) Every dealer shall maintain such books, accounts and records relating to his business as may be directed by the State Government.

(2) Every dealer shall submit monthly return relating to his business for the preceding month in Form ‘C’ to the licensing authority by the 5th day of every month.

21

National Seeds Policy, 2002

Aims and Objectives

It has become evident that in order to achieve the food production targets of the future, a major effort will be required to enhance the seed replacement rates of various crops. This would require a major increase in the production of quality seeds, in which the private sector is expected to play a major role. At the same time, private and Public Sector Seed Organisations at both Central and State levels, will be expected to adopt economic pricing policies which would seek to realise the true cost of production. The creation of a facilitative climate for growth of a competitive and localised seed industry, encouragement of import of useful germplasm, and boosting of exports are core elements of the agricultural strategy of the new millennium.

Biotechnology will be a key factor in agricultural development in the coming decades. Genetic engineering/modification techniques hold enormous promise in developing crop varieties with a higher level of tolerance to biotic and abiotic stresses. A conducive atmosphere for application of frontier sciences in varietal development and for enhanced investments in research and development is a pressing requirement. At the same time, concerns relating to possible harm to human and animal health and bio-safety, as well as interests of farmers, must be addressed.

Globalization and economic liberalization have opened up new opportunities as well as challenges. The main objectives of the National Seeds Policy, therefore, are the provision of an appropriate climate for the seed industry to utilize available and prospective opportunities, safeguarding of the interests of Indian farmers and the conservation of agro-biodiversity. While unnecessary regulation needs to be dismantled, it must be ensured that gullible farmers are not exploited by unscrupulous elements. A regulatory system of a new genre is, therefore, needed, which will encompass quality assurance mechanisms coupled with facilitation of a vibrant and responsible seed industry.

Thrust areas

1. Varietal development and plant variety protection

1.1. The development of new and improved varieties of plants and availability of such varieties to Indian farmers is of crucial importance for a sustained increase in agricultural productivity.

1.1.1. Appropriate policy framework and programmatic interventions will be adopted to stimulate varietal development in tune with market trends, scientific-technological advances, suitability for biotic and abiotic stresses, locational adaptability and farmers' needs.

1.2. An effective sui generis system for intellectual property protection will be implemented to stimulate investment in research and development of new plant varieties and to facilitate the growth of the Seed Industry in the country.

1.2.1. A Plant Varieties & Farmers' Rights Protection (PVP) Authority will be established which will undertake registration of extant and new plant varieties through the Plant Varieties Registry on the basis of varietal characteristics.

1.2.2. The registration of new plant varieties by the PVP Authority will be based on the criteria of novelty, distinctiveness, uniformity and stability.

1.2.3. The criteria of distinctiveness, uniformity and stability could be relaxed for registration of extant varieties, which will be done within a specified period to be decided by the PVP Authority.

1.2.4. Registration of all plant genera or species as notified by the Authority will be done in a phased manner.

1.2.5. The PVP Authority will develop characterisation and documentation of plant varieties registered under the PVP Act and cataloguing facilities for all varieties of plants.

1.3. The rights of farmers to save, use, exchange, share or sell farm produce of all varieties will be protected, with the proviso that farmers shall not be entitled to sell branded seed of a protected variety under the brand name.

1.4. The rights of researchers to use the seed/planting material of protected varieties for bonafide research and breeding of new plant varieties will be ensured.

1.5. Equitable sharing of benefit arising out of the use of plant genetic resources that may accrue to a breeder from commercialisation of seeds/planting materials of a new variety, will be provided.

1.6. Farmers/groups of farmers/village communities will be rewarded suitably for their significant contribution in evolution of a plant variety subject to registration. The contribution of traditional knowledge in agriculture needs to be highlighted through suitable mechanisms and incentives.

1.7. A National Gene Fund will be established for implementation of the benefit sharing arrangement, and payment of compensation to village communities for their contribution to the development and conservation of plant genetic resources and also to promote conservation and sustainable use of genetic resources. Suitable systems will be worked out to identify the contributions from traditional knowledge and heritage.

1.8. Plant Genetic Resources for Food and Agriculture Crops will be permitted to be accessed by Research Organisations and Seed Companies from public collections as per the provisions of the 'Material Transfer Agreement' of the International Treaty on Plant Genetic Resources and the Biological Diversity Bill.

1.9. Regular interaction amongst the Private and Public Researchers, Seed Companies/Organisations and Development Agencies will be fostered to develop and promote growth of a healthy seed industry in the country.

1.10. To keep abreast of global developments in the field of Plant Variety Protection and for technical collaboration, India may consider joining Regional and International Organisations.

1.11. The PVP Authority may, if required, resort to compulsory licensing of a protected variety in public interest on the ground that requirements of the farming community for seeds and propagating material of a variety are not being met or that the production of the seeds or planting material of the protected variety is not being facilitated to the fullest possible extent.

2. Seed production

2.1. To meet the Nation's food security needs, it is important to make available to Indian farmers a wide range of seeds of superior quality, in adequate quantity on a timely basis. Public Sector Seed Institutions will be encouraged to enhance production of seed towards meeting the objective of food and nutritional security.

2.2. The Indian seed programme adheres to the limited three generation system of seed multiplication, namely, breeder, foundation and certified seed. Breeder seed is the progeny of nucleus seed.

2.2.1. Nucleus seed is the seed produced by the breeder to develop the particular variety and is directly used for multiplication as breeder seed.

2.2.2. Breeder seed is the seed material directly controlled by the originating or the sponsoring breeder or Institution for the initial and recurring production of foundation seed.

2.2.3. Foundation seed is the progeny of breeder seed. Foundation seed may also be produced from foundation seed. Production of foundation seed stage-I and stage-II may thus be permitted, if supervised and approved by the Certification Agency and if the production process is so handled as to maintain specific genetic purity and identity.

2.2.4. Certified seed is the progeny of foundation seed or the progeny of certified seed. If the certified seed is the progeny of certified seed, then this reproduction will not exceed three generations beyond foundation stage-I and it will be ascertained by the Certification Agency that genetic identity and genetic purity has not been significantly altered.

2.3. Public Sector Seed Production Agencies will continue to have free access to breeder seed under the National Agriculture Research System. The State Farms Corporation of India and National Seeds Corporation will be restructured to make productive use of these organisations in the planned growth of the Seed Sector.

2.4. Private Seed Production Agencies will also have access to breeder seed subject to terms and conditions to be decided by Government of India.

2.5. State Agriculture Universities/ICAR Institutes will have the primary responsibility for production of breeder seed as per the requirements of the respective States.

2.6. Special attention will be given to the need to upgrade the quality of farmers' saved seeds through interventions such as the Seed Village Scheme.

2.7. Seed replacement rates will be raised progressively with the objective of expanding the use of quality seeds.

2.8. DAC, in consultation with ICAR and States, will prepare a National Seed Map to identify potential, alternative and non-traditional areas for seed production of specific crops.

2.9. To put in place an effective seed production programme, each State will undertake advance planning and prepare a perspective plan for seed production and distribution over a rolling (five to six year) period. Seed Banks will be set up in non-traditional areas to meet demands for seeds during natural calamities.

2.10. The 'Seed Village Scheme' will be promoted to facilitate production and timely availability of seed of desired crops/varieties at the local level. Special emphasis will be given to seed multiplication for building adequate stocks of certified/quality seeds by providing foundation seed to farmers.

2.11. For popularising newly developed varieties and promoting seed production of these varieties, seed minikits of pioneering seed varieties will be supplied to farmers. Seed exchange among farmers and seed producers will be encouraged to popularise new/non-traditional varieties.

2.12. Seeds of newly developed varieties must be made available to farmers with minimum time gap. Seed producing agencies will be encouraged to tie up with Research Institutions for popularization and commercialization of these varieties.

2.13. As hybrids have the potential to improve plant vigour and increase yield, support for production of hybrid seed will be provided.

2.14. Seed production will be extended to agro-climatic zones which are outside the traditional seed growing areas, in order to avoid unremunerative seed farming in unsuitable areas.

2.15. Seed Banks will be established for stocking specified quantities of seed of required crops/varieties for ensuring timely and adequate supply of seeds to farmers during adverse situations such as natural calamities, shortfalls in production, etc. Seed Banks will be suitably strengthened with cold storage and pest control facilities.

2.15.1. The storage of seed at the village level will be encouraged to facilitate immediate availability of seeds in the event of natural calamities and unforeseen situations. For the storage of seeds at farm level, scientific storage structures will be popularised and techniques of scientific storage of seeds will be promoted among farmers as an extension practice.

2.16. Seed growers will be encouraged to avail of Seed Crop Insurance to cover risk factors involved in production of seeds. The Seed Crop Insurance Scheme will be reviewed so as to provide effective risk cover to seed producers and will be extended to all traditional and non-traditional areas covered under the seed production programme.

3. Quality assurance

3.1. The Seeds Act will be revised to regulate the sale, import and export of seeds and planting materials of agriculture crops including fodder, green manure and horticulture and supply of quality seeds and planting materials to farmers throughout the country.

3.2. The National Seeds Board (NSB) will be established in place of existing Central Seed Committee and Central Seed Certification Board. The NSB will have permanent existence with the responsibility of executing and implementing the provisions of the Seeds Act and advising the Government on all matters relating to seed planning and development. The NSB will function as the apex body in the seed sector.

3.2.1. All varieties, both domestic and imported varieties, that are placed on the market for sale and distribution of seeds and planting materials will be registered under the Seeds Act. However, for vegetable and ornamental crops a simple system of varietal registration based on "breeders declaration" will be adopted.

3.2.2. The Board will undertake registration of kinds/varieties of seeds that are to be offered for sale in the market, on the basis of identified parameters for establishing Value for Cultivation and Usage (VCU) through testing/trialling.

3.2.3. Registration of varieties will be granted for a fixed period on the basis of multilocational trials to determine VCU over a minimum period of three seasons, or as otherwise prescribed as in the case of long duration crops and horticultural crops. Samples of the material for registration will be sent to the NBPGR for retention in the National Gene Bank.

3.2.4. Varieties that are in the market at the time of coming into force of the revised Seeds Act, will have to be registered within a fixed time period, and subjected to such testing as will be notified.

3.2.5. The NSB will accredit ICAR, SAUs, public/private organisations to conduct VCU trials of all varieties for the purpose of registration as per prescribed standards.

3.2.6. The NSB will maintain the National Seeds Register containing details of varieties that are registered. This will help the Board to coordinate and assist activities of the States in their efforts to provide quality seeds to farmers.

3.2.7. The NSB will prescribe minimum standards (of germination, genetic characteristics, physical purity, seed health, etc.) as well as suitable guidelines for registration of seed and planting materials.

3.2.8. Provisional registration would be granted on the basis of information filed by the applicant relating to trials over one season to tide over the stipulation of testing over three seasons before the grant of registration.

3.3. Government will have the right to exclude certain kinds or varieties from registration to protect public order or human, animal and plant life and health, or to avoid serious prejudice to the environment.

3.4. The NSB will have the power to cancel the registration granted to a variety if the registration has been obtained by misrepresentation or concealment of essential data, the variety is obsolete and has outlived its utility and if the prevention of commercial exploitation of such variety is necessary in the public interest.

3.5. Registration of Seed Processing Units will be required if such Units meet the prescribed minimum standards for processing the seed.

3.6. Seed Certification will continue to be voluntary. The Certification tag/label will provide an assurance of quality to the farmer.

3.6.1. The Board will accredit individuals or organisations to carry out seed certification including self-certification on fulfillment of criteria as prescribed.

3.7. To meet quality assurance requirements for export of seeds, Seed Testing facilities will be established in conformity with ISTA and OECD seed certification programmes.

3.8. The State Government, in conformity with guidelines and standards specified by the Board, will establish one or more State Seed Testing Laboratories or declare any Seed Testing Laboratory in the Government or non-Government Sector as a State Seed Testing Laboratory where analysis of seeds will be carried out in the prescribed manner.

3.9. Farmers will be encouraged to use certified seeds to ensure improved performance and output.

3.10. Farmers will retain their right to save, use, exchange, share or sell their farm seeds and planting materials without any restriction. They will be free to sell their seed on their own premises or in the local market without any hindrance provided that the seed is not branded. Farmers' right to continue using the varieties of their choice will not be infringed by the system of compulsory registration.

3.11. Stringent measures would be taken to ensure the availability of high quality of seeds and check the sale of spurious or misbranded seeds.

4. Seed distribution and marketing

4.1. The availability of high quality seeds to farmers through an improved distribution system and efficient marketing set-up will be ensured to facilitate greater security of seed supply.

4.2. For promoting efficient and timely distribution and marketing of seed throughout the country, a supportive environment will be provided to encourage expansion of the role of the private seed sector. Efforts will be made to achieve better coordination between State Governments to facilitate free Inter-State movement of seed and planting material through exemption of duties and taxes.

4.3. Private Seed Sector will be encouraged and motivated to restructure and reorient their activities to cater to non-traditional areas.

4.4. A mechanism will be established for collection and dissemination of market intelligence regarding preference of consumers and farmers.

4.5. A National Seed Grid will be established as a data-base for monitoring of information on requirement of seed, its production, distribution and preference of farmers on a district-wise basis.

4.6.Access to term finance from Commercial Banks will be facilitated for developing efficient seed distribution and marketing facilities for growth of the seed sector.

4.7. Distribution and marketing of seed of any variety, for the purpose of sowing and planting will be allowed only if the said variety has been registered by the National Seeds Board.

4.8. National Seeds Board can direct a dealer to sell or distribute seeds in a specified manner in a specified area if it is considered necessary to the public interest.

5. Infrastructure facilities

5.1.To meet the enhanced requirement of quality/certified seeds, creation of new infrastructure facilities along with strengthening of existing facilities, will be promoted.

5.2. National Seed Research and Training Center will be set up to impart training and build a knowledge base in various disciplines of the seed sector.

5.3.The Central Seed Testing Laboratory will be established at the National Seed Research and Training Center to perform referral and other functions as required under the Seeds Act.

5.4. Seed processing capacity will be augmented to meet the enhanced requirement of quality seed.

5.5.Modernisation of seed processing facilities will be encouraged in terms of modern equipment and latest techniques, such as seed treatment for enhancement of performance of seed, etc.

5.6. Conditioned storage for breeder and foundation seed and aerated storage for certified seed would be created in different regions.

5.7. A computerized National Seeds Grid will be established to provide information on availability of different varieties of seeds with production agencies, their location, quality etc. This network will facilitate optimum utilisation of available seeds in every region.

5.7.1. Initially, seed production agencies in the public sector would be connected with the National Seed Grid, but progressively the private sector will be encouraged to join the Grid for providing a clear assessment of demand and supply of seeds.

5.8. State Governments, or the National Seeds Board in consultation with the concerned State Government, may establish Seed Certification Agencies.

5.9.State Governments will establish appropriate systems for effective execution and implementation of the objectives and provisions of the Seeds Act.

6. Transgenic plant varieties

6.1. Biotechnology will play a vital role in the development of the agriculture sector. This technology can be used not only to develop new crops/varieties, which are tolerant to disease, pests and abiotic stresses, but also to improve productivity and nutritional quality of food.

6.2. All genetically engineered crops/varieties will be tested for environment and bio-safety before their commercial release, as per the regulations and guidelines of the Environment Protection Act (EPA), 1986.

6.3. The EPA, 1986, read with the Rules, 1989 would adequately address the safety aspects of transgenic seeds/planting materials. A list will be generated from Indian experience of transgenic cultivars that could be rated as environmentally safe.

6.4. Seeds of transgenic plant varieties for research purposes will be imported only through the National Bureau of Plant Genetic Resources (NBPGR) as per the EPA, 1986.

6.5. Transgenic crops/varieties will be tested to determine their agronomic value for at least two seasons under the All India Coordinated Project Trials of ICAR, in coordination with the tests for environment and bio-safety clearance as per the EPA before any variety is commercially released in the market.

6.6. After the transgenic plant variety is commercially released, its seed will be registered and marketed in the country as per the provisions of the Seeds Act.

6.7. After commercial release of a transgenic plant variety, its performance in the field, will be monitored for at least 3 to 5 years by the Ministry of Agriculture and State Departments of Agriculture.

6.8. Transgenic varieties can be protected under the PVP legislation in the same manner as non-transgenic varieties after their release for commercial cultivation.

6.9 All seeds imported into the country will be required to be accompanied by a certificate from the Competent Authority of the exporting country regarding their transgenic character or otherwise.

6.9.1. If the seed or planting material is a product of transgenic manipulation, it will be allowed to be imported only with the approval of the Genetic Engineering Approval Committee (GEAC), set up under the EPA, 1986.

6.10. Packages containing transgenic seeds/planting materials, if and when placed on sale, will carry a label indicating their transgenic nature. The specific characteristics including the agronomic/yield benefits, names of the transgenes and any relevant information shall also be indicated on the label.

6.11. Emphasis will be placed on the development of infrastructure for the testing, identification and evaluation of transgenic planting materials in the country.

7. Import of seeds and planting material

7.1. The objective of the import policy is to provide the best planting material available anywhere in the world to Indian farmers, to increase productivity, farm income and export earnings, while ensuring that there is no deleterious effect on environment, health and bio-safety.

7.1.1. While importing seeds and planting material, care will be taken to ensure that there is absolutely no compromise on the requirements under prevailing plant quarantine procedures, so as to prevent entry into the country of exotic pests, diseases and weeds detrimental to Indian agriculture.

7.1.2. All imports of seeds will require a permit granted by the Plant Protection Advisor to the Government of India, which will be issued within the minimum possible time frame.

7.2. All import of seeds and planting materials, etc. will be allowed freely subject to EXIM Policy guidelines and the requirements of the Plants, Fruits and Seeds (Regulation of import into India) Order, 1989 as amended from time to time. Import of parental lines of newly developed varieties will also be encouraged.

7.3. Seeds and planting materials imported for sale into the country will have to meet minimum seed standards of seed health, germination, genetic and physical purity as prescribed.

7.4. All importers will make available a small sample of the imported seed to the Gene Bank maintained by NBPGR.

7.5. The existing policy, which permits free import of seeds of vegetables, flowers and ornamental plants, cuttings, saplings of flowers, tubers and bulbs of flowers by certain specified categories of importers will continue. Tubers and bulbs of flowers will be subjected to post-entry quarantine.

7.5.1. After the arrival of consignments at the port of entry, quarantine checks would be undertaken; which may include visual inspection, laboratory inspection, fumigation and grow-out tests. For the purpose of these checks, samples will be drawn and the tests will be conducted concurrently.

8. Export of seeds

8.1. Given the diversity of agro-climatic conditions, strong seed production infrastructure and market opportunities, India holds significant promise for export of seeds.

8.2. Government will evolve a long term policy for export of seeds with a view to raise India's share of global seed export from the present level of less than 1% to 10% by the year 2020.

8.2.1. The export policy will specifically encourage custom production of seeds for export and will be based on long term perspective, dispensing with case to case consideration of proposals.

8.3. Establishment and strengthening of Seeds Export Promotion Zones with special incentives from the Government will be facilitated.

8.4. A data bank will be created to provide information on the International Market and on export potential of Indian varieties in different parts of the world.

8.5. A data base on availability of seeds of different crops to assess impact of exports on domestic availability of seeds will be created.

8.6. Promotional programmes to improve the quality of Indian seeds to enhance its acceptability in the International Market will be taken up.

8.6.1. Testing and certification facilities will be established in conformity with international requirements.

9. Promotion of domestic seed industry

9.1. Incentives will be provided to the domestic seed industry to enable it to produce seeds of high yielding varieties and hybrid seeds at a faster pace to meet the challenges of domestic requirements.

9.2. Seed Industry will be provided with a congenial and liberalized climate for increasing seed production and marketing, both domestic and international.

9.3. Membership to International Organisations and Seed Associations like ISTA, OECD, UPOV, ASSINSEL, WIPO, at the National level or at the level of individual seed producing agencies, will be encouraged.

9.4. Emphasis will be given to improving the quality of seed produced and special efforts will be directed towards improving the quality of farmers' saved seeds.

9.5. Financial support for capital investment, working capital and infrastructure strengthening will be facilitated through NABARD/ Commercial Banks/ Cooperative Banks.

9.6. Tax rebate/concessions will be considered on the expenditure incurred on in-house research and development of new varieties and other seed related research aspects. In order to develop a competitive seed market, the States will be encouraged to remove unnecessary local taxation on sales of seeds.

9.7. To encourage seed production in non-traditional areas including backward areas, special incentives such as transport subsidy will be provided to seed producing agencies operating in these marginalised areas.

9.8 Reduction of import duty will be considered on machines and equipment used for seed production and processing which are otherwise not manufactured in the country.

10. Strengthening of monitoring system

10.1. The Department of Agriculture & Cooperation (DAC) will supervise the overall implementation and monitoring of the National Seeds Policy.

10.2. The physical infrastructure in terms of office automation, communication facilities, etc., in DAC will be augmented in a time bound manner.

10.3. The technical capacity of DAC need to be augmented and strengthened to undertake the additional work relating to implementation of National Seeds Policy, implementation of PVP&FR Bill, Seeds Act, Import and Export of Seeds, etc.

10.4. Capacity building, including National and International training and participation in Seminars/Workshops will be organized for concerned officials.

11. Conclusion

The Government of India trusts that the National Seeds Policy will receive the fullest support of State Governments/Union Territory Administrations, State Agricultural Universities, plant breeders, seed producers, the seed industry and all other stakeholders, so that it may serve as a catalyst to meet the objectives of sustainable development of agriculture, food and nutritional security for the population, and improved standards of living for farming communities.

The National Seeds Policy will be a vital instrument in attaining the objectives of doubling food production and making India hunger free. It is expected to provide the impetus for a new revolution in Indian agriculture, based on an efficient system for supply of seeds of the best quality to the cultivator.

The National Seeds Policy will lay the foundation for comprehensive reforms in the seed sector. Significant changes in the existing legislative framework will be effected accompanied by programmatic interventions. The Policy will also provide the parameters for the development of the seed sector in the Tenth and subsequent Plans. The progress of implementation of the Policy will be monitored by a High Level Review Committee.

22

The Seeds Bill, 2004

A bill to provide for regulating the quality of seeds for sale, import and export and to facilitate production and supply of seeds of quality and for matters connected therewith or incidental thereto. Be it enacted by Parliament in the Fifty-Fifth Year of the Republic of India as follows:

CHAPTER I - Preliminary

1. Short title, extent, application and commencement:

(1) This Act may be called the Seeds Act, 2004

(2) It extends to the whole of India

(3) Save as otherwise provided in this Act, it shall apply to

  (a) every dealer; and

  (b) every producer of seed except when the seed is produced by him for his own use and not for sale.

(4) It shall come into force on such date as the Central Government may, by notification, appoint.

2. Definitions

In this Act, unless the context otherwise requires,

(1) "Agriculture" includes horticulture, forestry and cultivation of plantation, medicinal and aromatic plants;

(2) "Central Seed Testing Laboratory" means the Central Seed Testing Laboratory established or declared as such under sub-section (1) of section 32;

(3) "Certification Agency" means an agency established under section 26 or accredited under section 27 or recognised under section 30;

(4) "Chairperson" means the Chairperson of the Committee;

(5) "Committee" means the Central Seed Committee constituted under sub-section (1) of section 3;

(6) "Container" means a box, bottle, casket, tin, barrel, case, receptacle, sack, bag, wrapper or other thing in which any article or thing is placed or packed;

(7) "Dealer" means a person who carries on the business of buying and selling, exporting, or importing seed, and includes an agent of a dealer;

(8) "Export" means taking out of India by land, sea or air;

(9) "Farmer" means any person who cultivates crops either by cultivating the land himself or through any other person but does not include any individual, company, trader or dealer who engages in the procurement and sale of seeds on a commercial basis;

(10) "Horticulture nursery" means any place where horticulture plants are, in the regular course of business, produced or propagated and sold for transplantation;

(11) "Import" means bringing into India by land, sea or air;

(12) "Kind" means one or more related species or sub-species of crop plants each individually or collectively known by one common name such as cabbage, maize, paddy and wheat;

(13) "Member" means a member of the Committee;

(14) "Misbranded" - A seed shall be deemed to be misbranded if-

(i) It is a substitute for, or resembles in a manner likely to deceive, another variety of seed under the name of which it is sold, and is not plainly and conspicuously labelled so as to indicate its true nature;

(ii) it is falsely stated to be the product of any place or country;

(iii) it is sold by a name which belongs to another kind or variety of seed;

(iv) false claims are made for it upon the label or otherwise;

(v) when sold in a package which has been sealed or prepared by, or at the instance, of the dealer and which bears his name and address, the contents of each package are not conspicuously and correctly stated on the outside thereof within the limits of variability prescribed under this Act;

(vi) the package containing it, or the label on the package bears any statement, design or device regarding the quality or the kind or variety of seed contained therein, which is false or misleading in any material particular or if the package is otherwise deceptive with respect to its contents;

(vii) it is not registered in the manner required by or under this Act;

(viii) its label contains any reference to registration other than the registration number;

(ix) its label does not contain a warning or caution which may be necessary, and sufficient, if complied with, to protect human, animal and plant life and health or to avoid serious prejudice to the environment;

(x) the package containing it or the label on the package bears the name of a fictitious individual or company as the dealer of the kind or variety; or

(xi) t is not labelled in accordance with the requirements of this Act or the rules made thereunder;

(15) "Notification" means a notification published in the Official Gazette;

(16) "Prescribed" means prescribed by rules made under this Act;

(17) "Producer" means a person, group of persons, firm or organisation who grows or organizes the production of seeds;

(18) "Registered kind or variety", in relation to any seed, means any kind, or variety thereof, registered under section 13;

(19) "Registration Sub-Committee" means the Registration Sub-Committee constituted under sub-section (1) of section 7;

(20) "Regulation" means a regulation made by the Committee under this Act;

(21) "Seed" means any type of living embryo or propagule capable of regeneration and giving rise to a plant of agriculture which is true to such type;

(22) "Seed Analyst" means a Seed Analyst appointed under section 33;

(23) "Seed Inspector" means a Seed Inspector appointed under section 34;

(24) "Seed processing" means the process by which seeds and planting materials are dried, threshed, shelled, ginned or delinted (in cotton), cleaned, graded or treated;

(25) "Spurious seed" means any seed which is not genuine or true to type;

(26) "State Government", in relation to a Union territory, means the administrator thereof;

(27) "State Seed Testing Laboratory", in relation to any State, means the State Seed Laboratory established or declared as such under sub-section (2) of section 32 for that State;

(28) "Transgenic variety" means seed or planting material synthesized or developed by modifying or altering the genetic composition by means of genetic engineering;

(29) "Variety" means a plant grouping except micro-organism within a single botanical taxon of the lowest known rank, which can be :

- (i) defined by the expression of the characteristics resulting from a given genotype of that plant grouping;
- (ii) distinguished from any other plant grouping by expression of at least one of the said characteristics; and
- (iii) considered as a unit with regard to its suitability for being propagated, which remains unchanged after such propagation, and includes propagating material of such variety, extant variety, transgenic variety, farmers' variety and essentially derived variety. Footnote: "essentially derived variety", in respect of a variety (the initial variety) shall be said to be essentially derived from such initial variety when it,
  - (a) is predominantly derived from such initial variety, or from a variety that itself is predominantly derived from such initial variety, while retaining the expression of the essential characteristics that result from the genotype or combination of genotypes of such initial variety;
  - (b) is clearly distinguishable from such initial variety; and
  - (c) conforms (except for the differences which result from the act of derivation) to such initial variety in the expression of the essential characteristics that result from the genotype or combination of genotypes of such initial variety; Extant variety - "extant variety" means a variety available in India which is-

- (a) notified under section 5 of the Seeds Act, 1966; or
- (b) farmers' variety as defined in PVP Act; or
- (c) a variety about which there is common knowledge; or
- (d) any other variety which is in public domain.

CHAPTER II - The central seed committee, registration and other sub-committees

3. Constitution of Central Seed Committee

(1) The Central Government shall, by notification, constitute, for the purpose of this Act, a Committee to be called the Central Seed Committee.

4. Composition of the Committee

(1) The Committee shall consist of a Chairperson, members, ex-officio and other members, to be nominated by the Central Government.

(2) The Secretary to the Government of India in the Department of Agriculture and Co-operation, Ministry of Agriculture, shall be Chairperson, ex officio.

(3) The Committee shall consist of the following members, ex officio namely:-

(i) the Agriculture Commissioner, Department of Agriculture and Co-operation, Government of India;

(ii) the Deputy Director General (Crop Sciences), Indian Council of Agricultural Research;

(iii) the Deputy Director General (Horticulture), Indian Council of Agricultural Research;

(iv) the Joint Secretary in charge of seeds in the Department of Agriculture and Co-operation, Government of India

(v) the Horticulture Commissioner, Department of Agriculture and Co-operation, Government of India;

(vi) a representative of the Department of Bio-technology, Government of India, not below the rank of Joint Secretary to the Government of India;

(vii) a representative of the Ministry of Environment and Forests, Government of India, not below the rank of Joint Secretary to the Government of India.

(4) The Committee shall consist of the following other members to be nominated by the Central Government, namely

(i) the Secretary (Agriculture) from five States, one each from three out of the five geographical zones of the country as mentioned in the Schedule on rotation basis;

(ii) Director, State Seed Certification Agency from one State which is not represented under clause (i);

(iii) Managing Director, State Seeds Corporation, from one State which is not represented under clause (i) or clause (ii);

(iv) two representatives of farmers;

(v) two representatives of seed industry;

(vi) two specialists or experts in the field of seed development.

(5) The Committee may associate with it, in such manner, on such terms and for such purposes as it may deem fit, any person whose assistance or advice it may desire in complying with any of the provisions of this Act, and a person so associated shall have the right to take part in the discussion of the Committee relevant to the purposes for which he has been associated, but shall not have the right to vote and shall be entitled to receive such allowances or fees as may be fixed by the Central Government.

(6) A Member nominated under sub-section (5) shall, unless his seat becomes vacant earlier by resignation, death or otherwise, be entitled to hold office for two years from the date of his nomination but shall be eligible for re-nomination provided that the said member shall hold office only for so long as he holds the appointment by virtue of which his nomination was made.

(7) Save as otherwise provided, the terms and conditions of service of the members shall be such as may be prescribed.

(8) A member other than an ex officio member may resign his office by giving notice in writing to the Central Government and on such resignation being accepted, he shall be deemed to have vacated his office.

(9) A person shall be disqualified for being nominated or appointed as a member if he-

(i) has been convicted and sentenced to imprisonment for an offence which, in the opinion of the Central Government, involves moral turpitude; or

(ii) is an undischarged in solvent; or

(iii) is of unsound mind and stands so declared by a competent court.

(10) No act or proceeding of the Committee shall become invalid merely by reason of :

(i) any vacancy therein, or any defect in the constitution thereof; or

(ii) any defect in the appointment of a person acting as the Chairperson or a member of the Committee; or

(iii) any irregularity in the procedure of the Committee not affecting the merits of the case.

(11) The Central Government may, at any time, remove from office any member other than member, ex-officio after giving him a reasonable opportunity of showing cause against the proposed removal.

5. Powers and functions of the Committee

The Committee shall be responsible for and shall have all the powers for the effective implementation of this Act and shall advise the Central Government and the State Governments on matters relating to

(a) seed programming and planning;

(b) seed development and production;

(c) export and import of seeds;

(d) standards for registration, certification and seed testing;

(e) seed registration and its enforcement;

(f) such other matters as may be specified by the Central Government

6. Powers of committee to specify minimum limits of germination, purity, seed health, etc.

The Committee may, by notification, specify

(a) the minimum limits of germination, genetic and physical purity, and seed health, with respect to any seed of any kind of variety;

(b) the mark or label to indicate that such seed conforms to the minimum limits of germination, genetic and physical purity, and seed health specified under clause (a), and other particulars, such as expected performance of the seed in accordance with the information provided by the producer under section 14 which such mark or label may contain.

7. Registration and other Sub-Committees of the Committee and their functions

(1) The Committee shall constitute a Sub-Committee to be called the Registration Sub-Committee consisting of a Chairman and such number of other members, to assist him in the discharge of the functions of the Committee, as may be prescribed.

(2) It shall be the duty of the Registration Sub-Committee :

(a) to register seeds of varieties after scrutinizing their claims as made in the application in such manner as may be prescribed;

(b) to perform such other functions as are assigned to it by the Committee.

(3) The Committee may appoint as many other Sub-Committees including a Sub-Committee on Seed Certification as it deems fit consisting wholly of the members of the Committee or wholly of other persons or partly of members of the Committee and partly of other persons as it thinks fit to exercise such powers and perform such duties as may be delegated to them.

8. Procedure of the Committee and its Sub-Committees.

The Committee may, subject to the previous approval of the Central Government, make regulations for the purpose of regulating its own procedure and the procedure of any Sub-Committee thereof.

9. Secretary and other officers of the Committee.

The Central Government shall :

(a) appoint a person to be the Secretary of the Committee; and

(b) provide the Committee with such technical and other officers and employees as may be necessary for the efficient performance of the functions of the Committee under this Act.

10. Meetings of the Committee.

(1) The Committee shall meet as and when necessary at such time and place and shall observe such procedure in regard to transaction of business at its meetings (including the quorum at meetings) as may be provided by regulations.

(2) The Chairperson or, in his absence, the Agricultural Commissioner or, in the absence of both the Chairperson and the Agriculture Commissioner, any member chosen by the members present from amongst themselves, shall preside at a meeting of the Committee.

(3) All questions at a meeting of the Committee shall be decided by a majority of votes of the members present and voting and in the case of an equality of votes, the Chairperson or, in his absence, the Agriculture Commissioner or, in the absence of both the Chairperson and the Agriculture

Commissioner the person presiding shall have and exercise a second or casting vote.

11. State Seed Committee

Every State Government shall establish a State Seed Committee to :

(a) advise the Committee on registration of regional or local seeds of any kind or variety;

(b) advise the State Government on registration of seed producing units, seed processing units, seed dealers and horticulture nurseries;

(c) maintain, in each district, a list of seed dealers, seed producers, seed processing units and horticulture nurseries;

(d) seek information from persons engaged in the production, supply, distribution, trade or commerce in seeds of any kind or variety regarding stocks, prices, sales and other information in the manner as may be prescribed;

(e) advise the State Government and the Committee on all matters arising out of the administration and implementation of this Act; and

(f) carry out other functions assigned to, by, or under this Act.

CHAPTER III - Registration of kinds and varieties of seeds, etc.

12. Maintenance of National Register of seeds of kinds and varieties

(1) For the purposes of this Act, a register of all kinds and varieties of seed to be called the National Register of Seeds shall be kept by the Registration Sub-Committee wherein all specifications, as may be prescribed, shall be maintained.

(2) Subject to the directions of the Committee, the Register shall be kept under the control and management of the Registration Sub-Committee.

(3) The Registration Sub-Committee shall, within such intervals and in such manner as it thinks appropriate, publish the list of kinds and varieties of seed which have been registered during that interval.

13. Registration of seeds of any kind or variety

(1) No seed of any kind or variety shall, for the purpose of sowing or planting by any person, be sold unless such seed is registered under sub-section

(2) by the Registration Sub-Committee in such manner as may be prescribed.

(2) Subject to the provisions of sections 14 and 15, the Registration Sub-Committee may register, or refuse any kind or variety of seed on the basis of information furnished by the producer on the results of multi locational trials for such period as may be prescribed to establish the performance of that seed.

(3) The Registration Sub-Committee may grant provisional registration as prescribed to the varieties of seeds which are available in the market on the date of commencement of this Act.

(4) Registration made under this Act shall be valid for a period of fifteen years in the case of annual and biennial crops, and eighteen years for long duration perennials.

(5) At the expiry of the period granted under sub-section (4), the kind or variety of seed may be re-registered for a like period by the Registration Sub-Committee on the basis of information furnished by the producer on the results of such trials as may be prescribed under sub-section (2) to re-establish performance of the kind or variety of seed.

(6) The Registration Sub-Committee shall have the power to issue such directions to protect the interests of a producer against any abusive act committed by any third party during the period between the date of filing of application for registration and the date of decision by the Committee on such application.

14. Procedure for registration

(1) Every application for registration under sub-section (1) section 13 shall be made in such form and contain such particulars and be accompanied by such fees as may be prescribed.

(2) On receipt of any such application for the registration of a kind or variety of seed, the Registration Sub-Committee may, after such enquiry as it deems fit and after satisfying itself that the kind or variety of seed to which the application relates conforms to the claims made by the importer or by the seller, as the case may be, as regards the efficacy of the kind or variety of seed and its safety to human beings and animals, register the kind or variety, as the case may, of the seed on such conditions as may be specified by it and allot a registration number there to and issue a certificate of registration.

(3) The Registration Sub-Committee may, having regard to the efficacy of the seeds and its safety to human beings and animals, vary the conditions subject to which a certificate of registration has been granted and may, for that purpose, require the certificate holder by notice in writing to deliver the certificate to it within such time as may be specified in the notice.

15. Special provision for registration of transgenic varieties

(1) Notwithstanding anything contained in section 14, no seed of any transgenic variety shall be registered unless the applicant has obtained clearance in respect of the same as required by or under the provisions of the Environment (Protection) Act, 1986: Provided that the Registration Sub-Committee may, subject to clearance under the said Act, grant provisional registration, for a period not exceeding two years on the basis of information furnished by the producer on the results of multi-locational trials in the prescribed manner.

(2) Save as otherwise provided in sub-section (1), the form and manner in which and procedure for registration of transgenic variety of seed and the fee payable thereto shall be the same as applicable in case of registration under section 14.

16. Cancellation of registration of seeds of kinds and varieties

(1) The Registration Sub-Committee may cancel any registration granted under section 13 or section 15 or any one or more of the following grounds, namely

(a) that the holder of the certificate has violated any of the terms and conditions of the registration; or

(b) that the registration has been obtained by misrepresentation or concealment of essential data; or

(c) that the variety is not performing in accordance with the information provided by the producer under sub-section (3) of section 14 or has become obsolete or has outlived its utility; or

(d) that prevention of commercial exploitation of such variety of seeds is necessary.

(i) in the public interest;

(ii) to protect public order or public morality; or

(iii) to protect human beings, animal and plant life and health to avoid serious prejudice to the environment.

(2) No order of cancellation of registration under this section shall be made unless the holder thereof or the affected person concerned has been given a reasonable opportunity of showing cause in respect of the grounds for such cancellation.

17. Notification of cancellation of registration of seeds of kinds and varieties

The Registration Sub-Committee shall notify the cancellation of registration of a seed of any kind or variety made under section 13 or any registration made under section 15 in the Official Gazette

18. Exclusion of certain kinds or varieties of seed from registration

(1) Notwithstanding anything contained in this Act, no registration of any kind or variety of seed shall be made under this Act, if prevention of commercial exploitation of such kind or variety is necessary to protect public order or public morality or human, animal or plant life and health, or to avoid serious prejudice to the environment.

(2) A kind or variety of seed containing any technology, which is harmful, or potentially harmful, shall not be registered. Explanation.-For the purposes of this sub-section, the expression "technology" includes genetic use restriction technology and terminator technology.

19. Evaluation of performance

The Committee may, for conducting trials to assess performance, accredit centers of the Indian Council of Agricultural Research, State Agricultural Universities and such other organizations fulfilling the eligibility requirements as may be prescribed, to conduct trials to evaluate the performance of any kind or variety of seed.

20. Compensation to farmer

Where the seed of any registered kind or variety is sold to a farmer, the producer, distributor or vendor, as the case may be, shall disclose the expected performance of such kind or variety to the farmer under given conditions, and if, such registered seed fails to provide the expected performance under such given conditions, the farmer may claim compensation from the producer, distributor or vendor under the Consumer Protection Act, 1986.

21. Seed producers and seed processing units to be registered

(1) No producer shall grow or organize the production of seed unless he is registered as such by the State Government under this Act.

(2) No person shall maintain a seed processing unit unless such unit is registered by the State Government under this Act.

(3) The State Government shall register a producer or seed processing unit if he or it meets the specifications prescribed by the Central Government in terms of infrastructure, equipment and qualified manpower.

(4) Every application for registration under sub-section (3) shall be made in such form and manner and shall be accompanied by such fee as may be prescribed.

(5) The State Government may, after making such enquiry and subject to such conditions as it thinks fit, grant a certificate for maintaining a seed producing or a seed processing unit in such form as may be prescribed.

(6) Every seed producing and processing units shall furnish periodic returns to the Seed Certification Agency in such form and at such time as may be prescribed on the quantity of seeds of different kinds or varieties processed by them.

(7) The State Government may, after giving the holder of certificate of registration under sub-section (1), or sub-section (2), as the case may be, suspend or cancel the registration if¾

(a) such registration has been obtained by misrepresentation as to a material particular relating to the specification in terms of infrastructure, equipment or availability of qualified manpower; or

(b) any of the provisions of this Act or the rules made thereunder has been contravened.

22. Seed dealers to be registered

(1) Every person who desires to carry on the business of selling, keeping for sale, offering to sell, bartering, import or export or otherwise supply any seed by himself, or by any other person on his behalf shall obtain a registration certificate as a dealer in seeds from the State Government.

(2) Every applicant for dealership under sub-section (1) shall be required to furnish information about seed stocks, sales and other related information as may be prescribed.

(3) Even application for registration under sub-section(1) shall be made in such form and manner and shall be accompanied by such fee as may be prescribed.

(4) The State Government may, after making such enquiry and subject to such conditions as it thinks fit, grant a certificate of registration as a dealer in seeds in such form as may be prescribed.

(5) Every dealer registered under this section shall furnish such information and returns regarding seed stocks, seed lots, expiry date of seed lots and other related information as may be prescribed to the State Government.

(6) The State Government may, after giving the dealer an opportunity of being heard, suspend or cancel a certificate granted under this Act if-

(a) such registration had been obtained by misrepresentation of any material fact;

(b) contravenes any of the provisions of this Act or the rules made thereunder.

23. Horticulture nursery to be registered.

(1) No person shall conduct or carry on the business of horticulture nursery unless such nursery is registered with the State Government.

(2) Every application for registration under sub-section (1) shall be made in such form and contain such particulars and shall be accompanied by such fee as may be prescribed.

24. Duties of registration holders of horticulture nursery

Every person who is a holder of a registration of a horticulture nursery under section 23 shall

(a) keep a complete record of the origin or source of every planting material and performance record of mother trees in the nursery;

(b) keep a layout plan showing the position of the root-stocks and scions used in raising the horticulture plants;

(c) keep a performance record of the mother trees in the nursery;

(d) keep the nursery plants as well as the parent trees used for the production or propagation of horticulture plants free from infectious or contagious insects, pests or diseases affecting plants.

(e) furnish such information to the State Government on the production, stocks, sales and prices of planting material in the nursery as may be prescribed.

CHAPTER IV - Regulation of sale of seed and seed certification agencies

25. Regulation of sale of seeds of registered kinds and varieties

No person shall himself, or by any other person on his behalf, carry on the business of selling, keeping for sale, offering to sell, bartering, import or export or otherwise supply any kind of seed of any registered kind or variety unless

(a) such seed is identifiable as to its kind or variety;

(b) such seed conforms to the minimum limit of germination and genetic, physical purity, seed health specified under clause (a) of section 6;

(c) the container of such seed bears in the prescribed manner, the mark or label bearing the correct particulars thereof, specified under clause (b) of section 6;

(d) the container of such seed, in the case of transgenic varieties, bears a declaration to this effect as specified in sub-clause (2) of section 15;

(e) he complies with such other requirements as may be prescribed.

26. State Seed Certification Agency

The Committee may, in consultation with the State Government, by notification, establish a State Seed Certification Agency for the State to carry out the functions entrusted to the State Seed Certification Agency by or under this Act:

27. Accreditation of Seed Certification Agencies

(1) The Committee may in consultation with the State Government and the State Seed Committee, accredit :

(a) organizations to carry out certification, on the fulfillment of such criteria, as may be prescribed, or

(b) individuals or seed producing organisations to carry out self certification, in such manner as may be prescribed.

(2) The accredited individuals and seed producing organisations shall be subject to such inspection and control of the Committee, the concerned State Government and State Seed Certification Agency, as may be prescribed.

(3) The accreditation may be withdrawn by the Committee, for reasons to be recorded in writing and after giving to the concerned organization or individual, as the case may be, a reasonable opportunity of being heard.

28. Grant of certificate by the State Seed Certification Agency

(1) Any person selling, keeping for sale, offering to sell, bartering or otherwise supplying any seed of any registered kind or variety may, if he desires to have such seed certified by the State Seed Certification Agency, apply to that Agency for the grant of a certificate for the purpose.

(2) Every application under sub-section (1) shall be made in such form, shall contain such particulars and shall be accompanied by such fee as may be prescribed.

(3) On receipt of an application under sub-section (1), the State Seed Certification Agency may, after such enquiry as it thinks fit and after satisfying itself that the seed to which the application relates conforms to the prescribed standards, grant a certificate in such form and on such conditions as may be prescribed: Provided that such standards shall not be lower than the minimum limit of germination, genetic and physical purity specified for that seed under clause (a) of section 6.

29. Revocation of certificate

If the State Seed Certification Agency is satisfied, either on a reference made to it in this behalf or otherwise, that –

(a) the certificate granted by it under section 28 has been obtained by misrepresentation as to an essential fact; or

(b) the holder of the certificate has, without reasonable cause, failed to comply with the conditions subject to which the certificate has been granted or has contravened any of the provisions of this Act or the rules made thereunder, then, without prejudice to any other penalty to which the holder of the certificate may be liable under this Act, the State Seed Certification Agency may, after giving the holder of the certificate an opportunity of showing cause, revoke the certificate.

30. Recognition of seed certification agencies in foreign countries

The Central Government may, on the recommendation of the Committee and by notification, recognise any seed certification agency established in any foreign country, for the purposes of this Act.

CHAPTER V - Appeals

31. Appeals

(1) Any person aggrieved by a decision of the Registration Sub-Committee under section 14, section 16 or section 27 or of the State Seed Certification Agency under section 28 or section 29 may, within thirty days from the date on which the decision is communicated to him prefer an appeal to such authority (hereinafter referred to as the appellate authority) as the Central Government may think fit to constitute: Provided that the appellate authority may entertain an appeal after the expiry of the said period of thirty days if it is satisfied that the appellant was prevented by sufficient cause from filing the appeal in time.

(2) An appellate authority shall consist of a single person or three persons as the Central Government may think fit, to be appointed by that Government.

(3) The form and manner in which an appeal may be preferred under sub-section (1), the fee payable for such appeal and the procedure to be followed by the appellate authority shall be such as may be prescribed.

(4) On receipt of an appeal preferred under sub-section (1), the appellate authority shall, after giving the appellant and the other party an opportunity of being heard, dispose of the appeal as expeditiously as possible.

CHAPTER VI - Seed analysis and seed testing

32. Central and State Seed Testing Laboratories

(1) The Central Government may, by notification, establish a Central Seed Testing Laboratory or declare any seed-testing laboratory as the Central Seed Testing Laboratory to carry out the functions entrusted to the Central Seed Testing Laboratory by or under this Act in the prescribed manner

(2) The State Government may, in consultation with the Committee, and by notification, establish one or more State Seed Testing Laboratories or declare any seed testing laboratory in the Government or non-Government sector as a State Seed Testing Laboratory where analysis of seed of any kind or variety shall be carried out under this Act in the prescribed manner.

(3) Every Seed Testing Laboratory referred to in sub-section (1) shall have as many Seed Analysts as the Central Government may consider necessary.

(4) Every Seed Testing Laboratory referred to in sub-section (2) shall have as many Seed Analysts as the State Government may consider necessary.

33. Seed Analysts.

(1) In case of the Central Seed Laboratory, the Central Government and in other cases the State Government may, by notification, appoint such persons as the Government thinks fit and having the prescribed qualifications to be Seed Analysts and define the local limits of their jurisdiction.

(2) Every Central Seed Testing Laboratory established or declared under sub-section (1) of section 32 and every State Seed Testing Laboratory established or declared under sub-section (2) of that section shall have as many Seed Analysts as the Central Government or the State Government, as the case may be, specify.

34. Seed Inspectors

(1) The State Government may, by notification, appoint such persons as it thinks fit, having the prescribed qualifications, to be Seed Inspectors and define the areas within which they shall exercise jurisdiction.

(2) Every Seed Inspector shall be subordinate to such authority as the State Government may specify in this behalf.

35. Powers of Seed Inspectors

(1) The Seed Inspector may

(a) take samples of any seed of any kind or variety from-

(i) any person selling such seed; or

(ii) any person who is in the course of conveying, delivering or preparing to deliver such seed to a purchaser or a consignee; or

(iii) a purchaser or a consignee after delivery of such seed to him;

(b) send such sample for analysis to the Seed Analyst of the area within which such sample has been taken;

(c) enter and search, at all reasonable times, with such assistance, if any, as he considers necessary, any place in which he has reason to believe that an offence under this Act has been or is being committed and order in writing the person in possession of any seed in respect of which the offence has been or is being committed, not to dispose of any stock of such seed for a specific period not exceeding thirty days or, unless the alleged offence is such that the defect may be removed by the possessor of the seed, seize the stock of such seed;

(d) examine any record, register, document or any other material object found in any place mentioned in clause (c) and seize the same if he has reason to believe that it may furnish evidence of the commission of an offence punishable under this Act; and

(e) exercise such other powers as may be necessary for carrying out the purposes of this Act or any rule or regulation made thereunder.

(2) The power conferred by this section includes the power to break-open any container in which any seed of any kind or variety may be contained or to break-open the door of any premises where any such seed may be kept for sale: Provided that the power to break-open the door shall be exercised only after the owner or any other person in occupation of the premises, if he is present therein, refuses to open the door on being called upon to do so.

(3) Where the Seed Inspector takes any action under clause (a) of sub-section (1), he shall, as far as possible, call not less than two persons to be present at the time when such action is taken and take their signatures on a memorandum to be prepared in such form and manner as may be prescribed.

(4) The provisions of the Code of Criminal Procedure, 1973, or in relation to the State of Jammu and Kashmir, the provisions of any corresponding law in force in that State, shall, so far as may be, apply to any search or seizure under this section as they apply to any search or seizure made under the authority of a warrant issued under section 94 of the said Code, or, as the case may be, under the corresponding provisions of the said law.

CHAPTER VII- Export and import of seeds

36. Import of seeds

(1) All import of seeds

(a) shall be subject to the provisions of the Plants, Fruits and Seeds (Regulation of Import into India) Order, 1989, or any corresponding order made under section 3 of the Destructive Insects and Pests Act, 1914;

(b) shall conform to minimum limits of germination, genetic and physical purity, and seed health as prescribed under section 6; and

(c) shall be subject to registration as may be granted on the basis of information furnished by the importer on the results of multi-locational trials for such period as may be prescribed to establish performance.

(2) The Central Government may, by notification, permit to import an unregistered variety in such quantity and subject to fulfilling such conditions as may be specified in that notification for research purposes.

37. Export of seeds

The Central Government may, on the advice of the Committee, restrict, by notification, the export of seeds of any kind or variety if it is deemed that such export may adversely affect the food security of the country, or if it is felt that the reasonable requirements of the public will not be met, or on such other grounds as may be prescribed.

CHAPTER VIII- Offences and Punishment

38. Offences and punishment

If any person

(a) contravenes any provision of this Act or any rule made thereunder; or

(b) imports, sells, stocks or exhibits for sale or barter; and or otherwise supplies any seed of any kind or variety deemed to be misbranded ; or

(c) imports, sells, stocks or exhibits for sale or barter, or otherwise supplies any seed of any kind or variety without a certificate of registration; or

(d) obstructs the Committee, Registration Sub-Committee or Seed Certification Agency or Seed Inspector or Seed Analyst or any other authority appointed or duly empowered under this Act in the exercise of its powers or discharge of their duties under this Act or the rules made thereunder, he shall, on conviction, be punishable – with fine which shall not be less than five thousand rupees but which may extend to twenty five thousand rupees.

(2) If any person sells any seed which does not conform to the standards of physical purity, germination or health or does not maintain any records required to be maintained under this Act or the rules made thereunder he shall, on conviction, be punishable with fine which shall not be less than five thousand rupees but which may extend to twenty- five thousand rupees.

(3) If any person furnishes any false information relating to the standards of genetic purity, misbrands any seed or supplies any spurious seed or spurious transgenic variety, sells any non-registered seeds he shall, on conviction be punishable with imprisonment for a term which may extend to six months or with fine which may extend to fifty thousand rupees or with both.

39. Forfeiture of property

When any person has been convicted under this Act for the contravention of any of the provisions of this Act or the rules made thereunder, the seed in respect of which the contravention has been committed shall be forfeited to the Central Government.

40. Offences by companies

(1) Where an offence under this Act has been committed by a company, every person who at the time the offence was committed was in charge of, and was responsible to the company for the conduct of the business of the company, as well as the company, shall be deemed to be guilty of the offence and shall be liable to be proceeded against and punished accordingly:

Provided that nothing contained in this sub-section shall render any such person liable to any punishment under this Act if he proves that the offence was committed without his knowledge and that he exercised all due diligence to prevent the commission of such offence.

(2) Notwithstanding anything contained in sub-section (1), where an offence under this Act has been committed by a company and it is proved that the offence has been committed with the consent or connivance of, or is attributable to any neglect on the part of, any director, manager, secretary or other officer of the company, such director, manager, secretary or other officer shall also be deemed to be guilty of that offence and shall be liable to be proceeded against and punished accordingly. Explanation. – For the purpose of this section,-

(a) "company" means any body corporate and includes a firm or other association of individuals; and

(b) "director", in relation to a firm, means a partner in the firm.

CHAPTER IX- Power of central government

41. Power of Central Government to give directions to the State Governments

The Central Government may give such directions to any State Governments as may appear to the Central Government to be necessary for carrying into execution in the State any of the provisions of this Act or of any rule made there under.

42. Power of Central Government to issue directions to the Committee

(1) Without prejudice to the foregoing provisions of this Act, the Committee shall, in the discharge of its functions and duties under this Act, be bound by such directions on questions of policy as the Central Government may give in writing to it from time to time.

(2) The decision of the Central Government whether a question is one of policy or not shall be final.

43. Exemption from registration

(1) Nothing in this Act shall restrict the right of the farmer to save, use, exchange, share or sell his farm seeds and planting material, except that he shall not sell such seed or planting material under a brand name or which does not conform to the minimum limit of germination, physical purity, genetic purity prescribed under clause (a) or clause (b) of section 6.

(2) The Central Government may, by notification, and subject to conditions, if any, as it may specify therein, exempt from all or any of the provisions of this Act or the rules made thereunder, any educational, scientific or research or extension organization.

CHAPTER X- Miscellaneous

44. Protection of action taken in good faith

No suit, prosecution or other legal proceeding shall lie against the Government or any person for anything which is in good faith done or intended to be done under this Act.

45. Power to remove difficulties

(1) If any difficulty arises in giving effect to the provisions of this Act, the Central Government may, by order published in the Official Gazette, make such provisions not inconsistent with the provisions of this Act as may appear to be necessary for removing the difficulty: Provided that no order shall be made under this section after the expiry of two years from the date of commencement of this Act.

(2) Every order made under sub-section (1) shall be laid before each House of Parliament.

46. Power of Central Government to make rules

(1) The Central Government may by notification, make rules to carry out the provisions of this Act.

(2) In particular and without prejudice to the generality of the foregoing power, such rules may provide for all or any of the following matters, namely:-

(a) the terms and conditions of service of members of the Committee under sub-section (7) of section 4;

(b) the matters to be specified under clause (f) of section 5;

(c) the functions of the registration sub-committee under sub-section (1) of section 7;

(d) the manner of scrutinizing applications under clause (a) of sub-section (2) of section 7;

(e) the specifications which shall be maintained in the National Register of Seeds of kinds or varieties under sub-section (1) of section 12;

(f) the manner of registration of seed of any kind or variety under sub-section (1) and (3) of section 13;

(g) the period for which multi-locational trials shall be conducted under sub-section (2) of section 13;

(h) the form of application and the particulars which should be furnished in such application under sub-section (1) of section 14;

(i) the eligibility requirement which an organization shall fulfill for accreditation under section 19;

(j) the specification required to be fulfilled for registration as a producer or seed producing unit under sub-section (3) of section 21;

(k) the form and manner in which an application for registration under sub-section (3) of section 21 shall be made and the fee with which such application shall be accompanied under sub-section (5) of said section 21;

(l) the form in which a certificate for maintaining a seed producing or seed processing unit may be granted under sub-section (5) of section 21;

(m) the form in which and time within which periodic returns shall be filled under sub-section (6) of section 21;

(n) the information which an application for dealership in seeds shall be furnished under sub-section (2) of section 22;

(o) the form and manner in which an application for registration as seed dealer under sub-section (1) of section 22 shall be made and the fee which shall accompany such application under sub-section (3) of that section;

(p) the form in which a certificate of registration as a dealer in seeds shall be granted under sub-section (4) of section 22;

(q) the information and return which a registered dealer shall furnish to the State Government under sub-section (5) of section 22;

(r) the form in which an application for registration of a horticulture nursery shall be made, the particulars which such application shall contain and fee which shall accompany such application under sub-section (2) of section 23;

(s) the information on production, stocks, sales and prices of planting material in a nursery shall be furnished to the State Government under section 24;

(t) the requirement which a person carrying on business of selling, etc. of any registered kind or variety of seeds shall comply with under clause (e) of section 25;

(u) the criteria to be fulfilled under clause (a) and the manner of carrying out self-certification under clause (b) of sub-section (1) of section 27;

(v) the inspection and control of the Committee, the concerned State Government and the State Seeds Certification Agency for accrediting individuals and seed producing organizations under sub-section (2) of section 27;

(w) the form of application and the particulars to be furnished in such application and the fee which shall accompany such application under sub-section (2) of section 28;

(x) the form in which and the conditions subject to which a certificate shall be granted under sub-section (3) of section 28;

(y) the form and manner in which an appeal shall be preferred and the fee which such appeal shall accompany under sub-section (3) of section 31;

(z) the manner in which a Central Seed Testing Laboratory established or declared under sub-section (1) of section 32 shall carry out its functions;

(za) the manner of carrying out analysis of seeds shall be made under sub-section (2) of section 32;

(zb) the qualifications which a person to be appointed as Seed Analysts shall possess under sub-section (1) of section 33;

(zc) the qualifications which a person to be appointed as Seed Inspector shall possess under sub-section (1) of section 34;

(zc) the form and manner in which the memorandum shall be prepared under sub-section (3) of section 35;

(zd) the grounds on which the Central Government may restrict export of seeds under section 37;

(ze) any other matter which is to be or may be prescribed.

47. Power of Committee to make regulations

(1) The Committee may, with the previous approval of the Central Government, by notification, make regulations not inconsistent with the provisions of this Act and the rules made thereunder, to provide for all matters for which provision is necessary or expedient for the purpose of giving effect to the provisions of this Act.

(2) In particular and without prejudice to the generality of the foregoing power, such regulations may provide for all or any of the following matters, namely:-

(a) the procedure for conduct of business to be transacted by the Committee or any Sub-Committee thereof under section 8;

(b) the procedure in regard to transaction of business at meetings of the Committee (including the quorum at meetings)under sub-section (1) of section 10.

48. Rules and regulations to be laid before Parliament

Every rule and every regulation made under this Act shall be laid as soon as may be after it is made, before each House of Parliament, while it is in session, for a total period of thirty days which may be comprised in one session or in two or more successive sessions, and if, before the expiry of the session immediately following the session or the successive sessions aforesaid, both Houses agree in making any modification in the rule or regulation or both Houses agree that the rule or regulation should not be made, the rule or regulation shall, thereafter, have effect only in such modified form or be of no effect, as the case may be; so, however, that any such modification or annulment shall be without prejudice to the validity of anything previously done under that rule or regulation.

49. On the commencement of this Act, the Seeds Act, 1966 shall stand repealed

Provided that such repeal shall not affect -

(a) the previous operation of the law so repealed or anything duly done or suffered thereunder; or

(b) any right, privilege, obligation or liability acquired, accrued or incurred under the law so repealed; or

(c) any penalty, forfeiture or punishment incurred in respect of any offence committed against the Act so repealed; or

Repeal and Savings

(1) any investigation, proceeding, legal proceeding or remedy in respect of any such right, privilege, obligation, liability, penalty, forfeiture or punishment as aforesaid; and any such investigation, proceedings, legal proceeding or remedy may be instituted, continued or enforced; any such penalty forfeiture or punishment may be imposed as if this Act had not been passed:

Provided further that, subject to the first proviso and any saving provisions made elsewhere in this Act anything done, any action taken, any rule made, any notifications or orders issued under the provisions of the Act so repealed shall, in so far as they are not inconsistent with the provisions of this Act, be deemed to have been done, taken, made or issued under the corresponding provisions of this Act, and shall continue to be in force accordingly, unless and until expressly or implied repealed by anything done, action taken, rules made or, notification or orders issued under this Act.

(2) Notwithstanding such repeals any kind or variety of seeds that has been notified under the law as so repealed shall be deemed to have been registered under this Act, and any seed certification agency established under section 18 of the Seeds Act, 1966 shall be deemed to have been established or recognized, as the case may be, under this Act.

23

Seed Marketing - Marketing Structure and Organization

Seed marketing is one of the most vital components of seed technology. Broadly, it includes such activities as production, processing, storage, quality control and marketing of seeds. In the narrow sense, however seed marketing refers to the actual acquisition and selling of packed seeds, intermediate storage, delivery and sales promotional activities. Seed marketing comprises the following:

1. Demand forecasts (assessment of effective demand)
2. Marketing structure
3. Arrangements for storage of seeds
4. Sales promotional activities
5. Post-sales service
6. Economics of seed production and seed pricing

Demand Forecasts

The assessment of effective seed requirements is critical to any planned seed programme. The underlying principle in making demand forecasts should be that the seed supply keeps pace with seed demand (both present and future) in terms of quantity, quality, price, place and time. The outcome of such an approach would be planned seed production and marketing. It would also avoid shortages and gluts and as well ensure stable prices and profits.

In making demand forecasts, the following factors must be considered carefully.

(a) Total cultivated acreage, seed rate, quality replacement period and assessment of total potential seed requirement of each of the important crops.

(b) Impact of extension efforts on the introduction of improved production techniques, and future plans for promotion.

(c) Current acreage under high yielding varieties and amount of seed sold in the last year.

(d) Cultivator preferences for varieties, package size, kind of packing, quality and price.

(e) Number and size of competitors.

(f) Kinds of publicity and sales promotion that are most effective.

(g) Climate of the area where seed is being marketed.

Assessment of potential effective seed demand of the market, based on total seed requirements is of very little value, since the demand for high quality seed normally exists for the crop area which is under good fertility and irrigated conditions. The requirements for the remaining crop area are covered by uncontrolled production material obtained from the previous crop production. Furthermore, experience shows that the varietal purity and the yield potential of high quality seed of the self-pollinating varieties can be maintained by farmers during the reproduction process, without significant deterioration for three to four generations. Therefore, individual farmers only need to replace seed of self pollinating varieties every third or fourth year. Thus, the demand for high quality seed of self-pollinated crops is normally not higher than 25 to 30 per cent of the total requirement for areas under irrigated, and high fertility conditions. However there could be some exceptions, *e.g.,* if the climate of the region is not suitable for retaining viability of seed from the previous crop production. The farmers of such areas, if properly trained, can buy high quality seed each year, even of self-pollinated crops.

A rather different approach must be taken in the marketing of hybrid seeds, in which case new seed is needed by the farmer each season. Although, the critical period may be rather difficult, the subsequent planning is easier, particularly after sale statistics are seen to point in a certain direction. The dealers need to make periodic surveys of the market areas, to determine market potential, at least one season in advance. Dealer advance orders should be treated as informational material to aid the production section in organizing an effective production programme. The dealer should, however, not to be held to the exact amounts ofsuch advance orders.

The uses of demand forecasts are many and varied. A reliable forecast is the sheet anchor of all planning in business. Long term demand indications, in terms of quality, quantity, prices and term demand indications, in term of quality, quantity, prices and locations, help to make an investment decision, that is how much to invest, where to locatethe production facility, and how to organize marketing. Intermediate-range demand forecast help to make decisions on action necessary to optimize profits by balancing production and sales. Uses of short-term demand forecast include production planning and scheduling, distribution planning and scheduling, determination of targets and quotas for dealers and salesmen, planned buying of inputs, preparation of cash flow budgets, preparation of overall budgets and profits and loss statements, modifications of prices, policies, etc.

Marketing Structure (Establishment of effective channel of seed distribution)

The key to success in seed marketing is the establishment of effective channel of distribution. The various channels through which seed can be marketed vary greatly according to the needs of the seed company.

Present status of seed distribution

The types of seed distribution systems in India are:

a) *Farmers of farmer distribution:* This is the traditional method, whereby farmers obtain their requirements from neighbours either on cash payment or on an exchange basis. No formal marketing organization is required for this type of distribution.

b) *Distribution by co-*operatives: This involves procurement of seeds by cooperatives and its subsequent distribution. The distribution of seeds through cooperatives has often been encouraged by the government through subsidies and guarantees.

c) *Distribution by Departments of Agriculture:* Seeds are purchased by the government, out of the government funds, and are distributed through District Agricultural Officers and Block Development Officers.

d) *Distribution of seeds by non-government or quasi-government agencies*: In this system, the seeds are distributed through a network of seed distributors and seed dealers. Both the Seed Review Team (1968) and the National Commission on Agriculture (1976) have recommended that the State governments should withdraw from the seed procurement and distribution fields in a phased manner, so as to be able to concentrate on their principal function of providing extension education in the use of high quality seeds for improving productivity. Thus, emphasis is on the establishment of a seed marketing network to replace the role of State governments, and to establish a system that will be adequate for the anticipated increase in seed demand. In this connection, it is considered necessary that a network of seed dealers should be established.

Marketing Organization

There are a number of possible ways in which a marketing network could be organized. The simplest and most efficient system is to establish a central marketing cell and regional offices in end-use areas. The retail sale could be organized either by appointing distributors/dealers such as private dealers, cooperatives, agro-sales service centres, etc., or by opening seed company/ corporation-owned sales points, or both, as illustrated in Figure

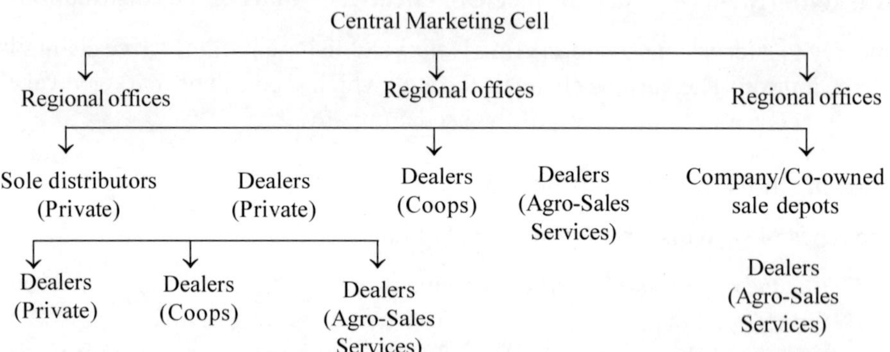

Under such a scheme, the central marketing cell is responsible for planning, appointment of dealers/distributors, seed movement, marketing intelligence research, pricing promotional activities, financing and record keeping. The regional offices are responsible for seed supply and promotional materials to dealers/distributors, training of seed dealers, expansion into new market areas, publicity and execution of promotional programmes. The dealers and distributors are the last, but most important link in the chain of seed marketing and they must assume great responsibilities. The responsibilities of dealers are,

1. Ethical dealings
2. Sell quality seed only
3. Proper storage of seeds
4. Maintenance of attractive shops
5. Wise use of publicity
6. Knowledge of product and competitors

The utmost care should be taken in the matter of appointment / selection of distributors and seed dealers. In this connection, due emphasis should be placed on an integrated marketing approach. The dealers involved in selling fertilizers

/ pesticides, or other agricultural inputs, should be preferred. The dealer distributor network should be organized in such a manner that the seed retailers are located within easy reach of farmers *e.g.*, establishment of at least four sales points in each Community Development Block.

References

Agrawal, R.L. 1995. *Seed technology*. Oxford and IBH Publishing. New Delhi.

Bhaskaran, M., K.Vanangamudi, A.Bharathi, P.Natesan, R.Jerlin, N.Natarajan and K.Prabhakaran. 2003. *Principles of Seed Production and Quality Control*. Kalyani Publishers, Ludhiana.

http://agricoop.nic.in/seedpolicy.html

http://agricoop.nic.in/seeds/seeds_bill.html

http://agricoop.nic.in/seedsact.html

http://en.wikipedia.org/wiki/Five-Year_Plans_of_India

http://en.wikipedia.org/wiki/Seed

http://nsrtc.nic.in/testing.html

http://plantquarantineindia.nic.in/PQISPub/html/seepol.html

http://www.agriinfo.in/?page=topic&superid=3&topicid=78

http://www.indiaagronet.com/indiaagronet/seeds/CONTENTS/general_pr...

http://www.mpkrishi.org/EngDocs/AgriLeft/alliedAgenciesSeedCerti.aspx

http://www.thefreedictionary.com/seed

http://seednet.gov.in/PDFFILES/Seed_Control_Order_1983.pdf

http://krishi.bih.nic.in/Acts-Rules/Seed_Rules_1968.pdfh

ttp://seednet.gov.in/PDFFILES/National%20Seed%20Policy,%202002.pdf